EMPOWER

HOW TO CO-CREATE THE FUTURE

Also by David Passiak

Disruption Revolution:
Innovation, Entrepreneurship, and
the New Rules of Leadership

Red Bull to Buddha:
Innovation and the Search for Wisdom

EMPOWER

HOW TO CO-CREATE THE FUTURE

DAVID PASSIAK

Social Meditate Press

Cover Design: Tom Lau

Printed in the United States of America

First Printing 2016

ISBN 978-0-9898233-2-6

Social Meditate Press
745 Telya Ridge
Milford, MI 48381
www.SocialMeditate.com

This book is dedicated to future generations

May you aspire to do better than your ancestors,
surpass the hopes and dreams captured in this book,
and empower humanity to reach our true potential

Contents

Co-Creation By Donation

My intention with *Empower* is to start a global conversation about how to co-create the future. The full version of the e-book is available by donation at CoCreateTheFuture.com Anyone can download it for free.

I believe equal access to knowledge and ideas is more important than pursuing profits, and a $15–$25 retail price is too expensive for millions of entrepreneurs, students, and people in opportunity markets like Eastern Europe, Africa, India or Southeast Asia with average salaries of $200–$400/month.

I am also a strong proponent of giving first, in the spirit of the collaborative and sharing economy. As a practitioner of meditation for 20 years and former scholar of religion, I consider my work part of a longstanding tradition to give teachings without expectation.

I greatly appreciate the generosity of all who can afford to give. Thanks for your support. I hope you enjoy the book!

With Gratitude,
David Passiak

Foreword

The emergence of new technologies is changing human behavior and causing rifts in society, business, and the economy. It is imperative that we pay attention to this rising movement and its ripple effects across the world, as it will impact each of our lives.

What is this movement? The Collaborative Economy is borne out of new Internet technologies and enables people to get what they need from each other. You've heard of young tech startups like Uber, Airbnb, BlaBlaCar, Instacart, Lending Club, Blockchain, and Kickstarter, all of which enables peer-to-peer commerce. They are displacing inefficient institutions like taxis, hotels, railroads, banks, credit systems, and financiers. Since this is a global movement, each region of the world sports its own versions: India has Ola, China has Didi, and the Arabian nations have their own versions. The people are even creating their own goods by using 3D printers, maker spaces, and online tools that help them learn, transact, and support these goods in the Maker Movement.

This movement shows no sign of slowing down, as our research has found that more than half of North Americans are participating, and we've read reports of growth across all regions, even with these startups being less than a decade old. The funding for this market is unprecedented; investors have deployed more than $30 billion in these startups, fueling a market economy that is quickly flourishing. Of course, this hasn't occurred without friction; incumbents are pushing back, and city and federal regulators are struggling to stay abreast of the rapid changes.

All of this is happening at a dizzying pace, and whether you work at a startup, company, nonprofit, or government agency, there are even more changes to come in the future. What the industry needs is a guide, and David Passiak provides just that.

David Passiak encourages us in *Empower* to think about the future like we are building a movement. Big companies and startups collaborate with empowered people. There are shared values, common goals, and purpose. David presents a series of conversations about collaboration and sharing from a variety of perspectives within an inner circle.

Who's in this inner circle that's now made available to you in this book? These perspectives on the Collaborative Economy range from its best-respected advocates, such as Robin Chase, Neal Gorenflo, Antonin Léonard, Arun Sundararajan, and Chelsea Rustrum, to likeminded business leaders and bestselling authors, such as Adam Grant, Brad Feld, Alex Bogusky, Shane Snow, and Rita Gunther McGrath. David makes all of this insight easily accessible to a general audience and groups people together.

I first met David Passiak three years ago when he interviewed me about the Collaborative Economy for his last book, *Disruption Revolution*. David gathered an impressive group of disruptive thinkers, including Seth Godin, Chris Anderson, Robert Scoble, Sarah Lacy, James Altucher, Erik Qualman, and Brian Solis. I liked that he took a wisdom-of-the-crowd approach to trends in innovation.

At the time in 2013, I was preparing to launch Crowd Companies, an innovation council to help large companies participate in the Collaborative Economy and Autonomous World. David and I were both seeing and articulating how these changes would impact the world.

Three years later, the Collaborative Economy and the Autonomous World are top of mind for many entrepreneurs and business leaders. Crowd Companies has grown internationally to include members like BMW, Cisco, MasterCard, Pepsi, Visa, Nestle, and more.

The pace of innovation will continue exponentially. The Collaborative Economy is positioned to become the world economy. David has once again gathered an impressive group of visionary leaders to help us understand what lies ahead. *Empower: How to Co-Create the Future* is a great starting point for entering the conversation.

—Jeremiah Owyang
Silicon Valley
November 2016

Acknowledgments

First and foremost, I am grateful to all of the participants in this book. Thanks for the precious gift of your time and for the honor to share our conversations with a wider audience. Thank you especially to Jeremiah Owyang for writing the foreword and encouraging me to take on another monumental book-length project.

Thanks also to my business partner and lifelong friend Marc Joseph, my literary agent and good friend Matthew Guma for his support and guidance, to Jessica Baker for her wonderful transcription work, and to Tom Lau for an incredibly awesome cover that captures the themes of collaboration and co-creation in a simple and powerful way.

I visited around 40 countries since my last book *Disruption Revolution* came out. That never would have happened if I didn't make it available by donation. There are too many people to thank by name along the way, but thanks to Alex Hutley and Arto Joensuu of Dubizzle for inviting me to join you in Dubai as your Head of Innovation and Research, which was the first stop on a magical journey.

I'm deeply humbled by the power of collaboration and sharing that makes it possible to do this work. May we continue to take risks and trust in the power of humanity. I honor and acknowledge the potential within us all.

Introduction

This is an incredible time to be alive. Not only are we living through the period of the most rapid innovation and exponential change in human history, but for the first time we have the tools and resources to solve the world's biggest problems. The collaborative economy and rise of the autonomous world will radically reconfigure business, society, and culture. Everything is going to change, faster than we think.

Humanity is building a bridge from an analog to digital world, taking a giant leap forward on par with the discovery of fire, invention of writing, or going to the moon. Each of us plays a role in co-creating the future. Entrepreneurs, investors, activists, employees, students, politicians, and community leaders must work together with passion and purpose. The leadership decisions we make on how to coexist with technology will shape the evolution of civilization for the next 1,000 years:

Will we live in abundance, leveraging the power of technology to share value and realize our true potential in an emerging networked society? Or will automation from AI, drones, robots, self-driving vehicles, and the Internet of Things (IoT) decimate jobs, lead to mass inequality and a dystopian nightmare?

Empower advocates an approach to co-creating the future akin to building a global movement. It is not enough to think about how to protect our individual jobs from automation or save our companies from disruption. If you are reading this book, then there is a higher likelihood that you have the skill sets and resilience to succeed in this new era, but millions of people will not. We need to think about how to connect our efforts to a greater purpose, shared goals, and a global community of like-minded individuals committed to building a collaborative society.

Movements require collaboration among members that mutually support and look out for each other. Their strong sense of community provides a safe space to share and test ideas, transforms failures into learning opportunities, and builds resilient networks that allow us to adapt and grow. Movements have visionary leaders that inspire us to embrace challenges and transcend differences for the sake of the greater good. They foster a sense of optimism and aim to accomplish what seems impossible.

In the context of movements, technology is a great enabler. It brings people together by allowing them to communicate better, forge new relationships and communities organized around like-minded interests, and facilitate exchanges of ideas, goods, and services. Early-adopters test new technologies and baptize innovations into popular culture through reviews, recommendations, blogs, and social media. Terms like "evangelism" and "conversion" borrowed from religious movements reflect the passion and purpose at the heart of every startup team. Technology provides access to knowledge and opportunities irrespective of race, religion, ethnicity, or gender.

The first step toward building a movement is to start a conversation around the issues and challenges that lie ahead. *Empower* takes a holistic and integrated approach covering: (1) Collaborative Leadership, (2) The Power of Sharing, (3) Companies and the Crowd, and (4) Better Economics. Each participant in the book is a recognized thought leader whose contributions help to advance a new way of being and conducting business. Because the full e-book is available by donation, anyone with access to a computer or handheld device can join the conversation.

Movements are my area of expertise. I have 20 years of professional and research experience building movements around brands, organizations, and startups. My industry experience ranges from leading social media for Volkswagen, where I launched and grew a community of tens of millions of fans across Facebook, YouTube, Twitter, and their first company blog; to a year in Dubai serving as Head of Innovation and Research for

Dubizzle, building a movement around a website that 40% of the country visited once/month; to writing three books *Red Bull to Buddha*, *Disruption Revolution,* and *Empower*. I have visited around 40 countries, speaking with innovators and entrepreneurs about this shared vision.

Prior to industry, I was a Ph.D. student at Princeton studying grassroots religious and cultural movements referred to as "Great Awakenings." My research covered the Sixties Counterculture and Civil Rights movements, which became templates for modern activism, community building, and word-of-mouth marketing. My M.A. thesis at Arizona State University explored the Great Awakenings of the 1700s and 1800s that led to democracy and the abolition of slavery. At the Tanenbaum Center for Interreligious Understanding, I ran the Religion and Conflict Resolution Program, promoting the work of religiously-motivated peacemakers that put their lives at risk to build nonviolent peace movements around the world.

We need a different mindset to co-create the future. This includes a new approach to leadership characterized by collaboration and sharing, cooperation and openness, transparency and honesty. We have to think holistically about the roles that our brands, organizations, or startups play in the world. New technologies and platforms generate new business models, create opportunities to share value, and allow us to rethink the basic building blocks of business, society, and culture.

DISRUPTIVE INNOVATION, THE COLLABORATIVE ECONOMY, AND THE RISE OF THE AUTONOMOUS WORLD

Empower is a follow-up to *Disruption Revolution*, a survival guide and innovation handbook on how to succeed when disruption is the new norm. After the economic crash of 2008–2009, I noticed a pattern of innovators and entrepreneurs rallying around the term "disruption." *TechCrunch* baptized the term with its first Disrupt conference in 2011. As the rest of the world struggled to get back to normal, billions of dollars in venture capital accelerated efforts to leapfrog ahead. Innovation across every industry gave rise to what I termed a "disruption revolution."

I decided the best way capture the strategies, best practices, and emerging trends in innovation was to interview the visionary thinkers creating them. *Disruption Revolution* includes over 20 leading innovators such as Seth Godin, Chris Anderson, Robert Scoble, Brian Solis, Sarah Lacy, James Altucher, Alex Osterwalder, Erik Qualman, and Jeremiah Owyang. The book is organized around an integrated and holistic innovation process covering: (1) Research and Trends, (2) Mission and Purpose, (3) Teams and Operations, and (4) Sales, Marketing and Communications. This parallels the same four-part framework of *Empower.*

Disruption Revolution was the first comprehensive and in-depth collection of perspectives on disruptive innovation. After release it "went viral" and caught the attention of Dubizzle, the sixth most visited site in the UAE and largest P2P marketplace in the Middle East. Within a week I had an offer to be Head of Innovation and Research and moved to Dubai. I introduced Jeremiah Owyang to the founders of STEP Conference. They invited him to give a keynote on the Collaborative Economy. Our conversations in Dubai led to the idea for *Empower.*

A lot has changed in three years since I published *Disruption Revolution.* The global economy recovered, though in many ways this bounce-back effect gave business leaders a false sense of confidence: The crash accelerated trends in innovation that will radically disrupt traditional industries. As Brian Solis said at the time, constraint drives innovation. Forced to do more with less resources, companies embraced new products, services, and business models. Many startups founded in the wake of the crash are now thriving. At the time of writing, there are 177 unicorns (privately held companies with over $1b valuation) with a cumulative valuation of $677b.

The collaborative and sharing economy is now mainstream, including a number of unicorns like Uber, Airbnb, Lyft, and WeWork. Empowered people can get whatever they need directly from each other. Growth in this arena outpaced the rise of social media and will continue exponentially. However, as Chelsea Rustrum points out, what began as a movement is being eaten by an on-demand everything model. People-powered platforms of the collaborative and sharing economy face disruption with the rise of the autonomous world (drones, robots, AI, self-driving vehicles, etc.) All of this innovation will bring a tsunami-sized wave of disruption. Every industry will require new business models and management processes, face huge regulatory challenges, and automation could lead to mass unemployment.

The traditional career path is officially over, replaced by the gig economy and project-based temporary workers competing with machines for jobs. A recent report indicated 83% of jobs making less than $20/hour are at risk of automation. This trend could be exacerbated by findings that millennials "don't like dealing with people." Instead, they prefer kiosks and automated services. White-collar, office jobs could also be replaced. In fact, they are often easier to automate because software and AI doesn't require physical installations. As Martin Ford puts it, any repetitive task can be automated.

The doom and gloom of techno-dystopia is counterbalanced by a profound shift in values and behavior. Millennials buy less stuff, prefer access to ownership, embrace collaboration and sharing, and want to support and work for companies that have a social purpose. New communities form around likeminded interests, challenging the importance of traditional identity categories of race, gender, ethnicity, nationality, and religion. The collaborative economy empowers people to create, share, and exchange goods and services directly from each other, eliminating the need for intermediaries.

We are also seeing widespread interest in meditation, mindfulness, and awakening. This is accompanied by a new generation of apps, devices, and tools to help relax and attain altered states of consciousness. What my meditation teacher Kenneth Folk calls "contemplative fitness" could be incredibly important as we need to train our minds for immersion in a world saturated by virtual and augmented reality. The giant leap we are making with technology may be a catalyst for the next Great Awakening. This was a main theme in my first book *Red Bull to Buddha* and will be the subject of a future project called *Technologies of the Soul*.

The next generation of collaborative economy startups can leverage technology to distribute and share value with users. Decentralized platforms built on the blockchain could enable new types of company creation and eliminate the need for hierarchical power structures. Digital cooperatives and open innovation models present opportunities to reorganize society around communal interests. The world is getting more interconnected, giving rise to what Parag Khanna calls an "emerging networked society."

As Brad Feld points out in his interview on giving first, part of the challenge we face is that humans are very bad at predicting outcomes of exponential technologies. Cycles of rapid change and adoption that previously took decades to unfold can now happen in a matter of weeks or months. The ripple effects from breakthroughs in innovation become more complicated when factoring in the idea of AI learning from AI. We are tasked with preparing for the future when even the best minds in the world can't predict what it will be like 10–20 years from now.

Equally difficult to predict are the changing dynamics of global communities. Current economic leaders like the U.S. and Europe face vast challenges, plagued with regulatory red tape, legacy infrastructure, massive organizations that will need to be restructured, and millions of jobs at risk of automation. Meanwhile emerging markets across South America, Africa, and Asia present greenfield opportunities to improve the quality of life for billions of people coming online for the first time. Access to the same tools and platforms could allow developing nations and megacities to leapfrog ahead, changing the global flow of people, money, and power.

Recent history has shown that underemployment and social media can be a recipe for nationalism and religious extremism. The same platforms that connect billions of people through collaboration and sharing, community and open dialogue, serve as amplifiers and echo chambers of polarizing anger and hatred. Attempts to enforce isolationist immigration policies distract from the ominous threat of AI replacing jobs, while the rise of the autonomous world presents new security risks as self-driving vehicles, drones, and robots could all be used as weapons.

The interrelationship between technology and humanity, innovation and tradition, globalization and local identities, community and the individual, is vast and complex. Although we cannot control the rapid pace of innovation, we can choose to face the future with a sense of hope and optimism, mindfulness and compassion. We can be part of the problem, or part of the solution.

EMPOWER: BUILDING A MOVEMENT TO CO-CREATE THE FUTURE

If *Disruption Revolution* attempted to answer the question, "How do we innovate?" then *Empower* asks, "How do we co-create the future we want to live in?"

The short answer and main idea throughout the book is to approach the future like we are building a movement. Movements follow predictable patterns. First, the founders have an idea to bring about some type of change in the world. Next, they form a team or core group around a shared mission and purpose. The team then evangelizes their product, service, or message with friends and family, who share with their friends and family, and so on to spread by word-of-mouth. As this process continues, the message and story become crystalized, allowing the movement to build momentum and scale, taking on increasingly more ambitious goals. *Empower* is organized around these predictable patterns, creating a holistic framework for everyone to join the conversation.

Part I - Collaborative Leadership Adam Grant shares insights on how givers command respect, encourage collaboration, and can be more successful leaders based on his bestselling books *Give and Take* and *Originals*. Robin Chase presents her *Peers Inc.* model for companies of the future that combine the power of platforms and the crowd (Peers), with the scale of industrial production (Inc.). Brad Feld speaks about his philosophy to #GiveFirst and the power of networks in startups and startup communities. We end with Douglas Atkin detailing how to build movements based on his experience as Global Head of Community for Airbnb and author of *The Culting of Brands*.

Part II - The Power of Sharing presents a vision for a people-powered movement where startups, companies, nonprofits, and governments work together to co-create a more prosperous society.

It opens with Neal Gorenflo's inspiring personal journey going from a corporate executive to founder of Shareable, a catalyst for thousands of sharing initiatives around the world. Next we speak with Prince EA about his message of love and hope that organically built a platform of six million fans and a billion video views. Antonin Léonard shares lessons learned from co-founding OuiShare, a decentralized global organization working towards building a collaborative society. Chelsea Rustrum reminds us why we should resist reducing the sharing economy to an on-demand everything model, and instead focus on how technology can serve and empower us to live in abundance.

Part III - Companies and the Crowd looks at the leadership challenges faced by people working in larger organizations to tap into and align with the power of the crowd. Shane Snow shares insights from building Contently, a leading crowd company with a community of 55,000 journalists servicing some of the biggest brands in the world. Michael Bronner explains why the essence of strategy is sacrifice and how leaders could be more successful if they focused on consciousness. Rita Gunther McGrath presents a new model for strategy where companies compete in arenas, not marketplaces, due to the end of sustainable competitive advantage. And last Alex Bogusky, Creative Director of the Decade for 2000–2010, tells us about the future of brands and why consumers get what they allow, not what they deserve.

Part IV - The Economy of the Future looks at wide-scale, macro-economic trends emerging from the collaborative economy and rise of the autonomous world. Arun Sundararajan opens with a survey of ideas on "crowd-based capitalism" from his latest book *The Sharing Economy*. Douglas Rushkoff explains how we can stop being trapped in an unsustainable economic model based on unlimited growth. Martin Ford advocates a public conversation about the risk of widespread unemployment due to AI based on his bestselling book *Rise of the Robots*. The book closes with a vision of an emerging networked society from Parag Khanna's groundbreaking book *Connectography*.

I hope you are challenged and inspired as much as I am by these incredible conversations with some of the world's most visionary thinkers and change agents. Each interview required extensive research, preparation, and editing in an effort to present the first comprehensive book of its kind on the challenges and opportunities we face as a society. Together we can empower each other to reach our true potential and co-create the future we want and deserve.

PART I

COLLABORATIVE LEADERSHIP

Why does success come from empowering others?

How can you build a movement around your organization?

Why are networks a leader's greatest asset?

GIVING AND ORIGINALITY

Adam Grant

Adam Grant demonstrates in *Give and Take* that giving and collaboration can lead to greater success than acting out of greed and self-interest. This is especially true when networks and social media make our actions more transparent and the world of work is more relational. His latest book *Originals* builds upon this foundation by teaching people to champion new ideas in the workplace:

- Authenticity can be more powerful than charisma
- Why givers make great leaders
- Who you know matters as much as how you work
- Having a meaningful impact increases productivity

Adam's passion for sharing comes through in the first moments of our conversation. His commonsense examples and thought-provoking research break down conventions to illuminate a new framework for leadership. It is easy to see why he is the top-rated professor at Wharton for five years and counting. I can't wait for his next book, which will be co-authored with Sheryl Sandberg. This is a fun and engaging read.

DP: I love "a-ha" moments where something unexpected leads to profound insights. One of my favorite stories of yours is from a call center experience, where a five-minute speech by a very quiet woman led to a 400% increase in results from the entire team. Can you tell us about that experience and how it helped you to understand the importance of giving in the workplace?

AG: I was in this call center working on a research project for my Ph.D. University fundraisers were calling alumni for donations. We had a massive amount of turnover. A lot of the callers described the job as repetitive and demotivating. They complained that they were yelled at all the time or rejected very frequently. My thought was if the callers knew how their work was making a difference, it might make it more meaningful and motivating.

We had a couple of managers tell their best stories about how the money was being used, and we saw no impact whatsoever. We realized it was the right message, but the wrong story. Instead of being motivated secondhand by managers who had an ulterior motive of getting you to work harder, we should invite people who were being served and helped by the work that the callers were doing.

We brought in a scholarship student who talked about how the money raised by callers enabled him to go to school, and how much he appreciated the work that the callers were doing. The average caller spiked 142% in weekly phone minutes and 171% in weekly revenue. The student was really charismatic. He had been named "Most Likely to Become President" out of his college class.

I was like, "Ok, is this the Scholarship Student Effect or is this the Charisma Effect?" I tried to find the opposite of the charismatic student, and that was Emily. She was an extremely quiet, shy student who basically looked at her feet the entire time. I was worried that none of the callers were going to be inspired by her, but the Emily Effect ended up being about two and a half times as strong. The average caller spiked at more than 400% in weekly revenue.

I think it was easier to empathize with someone who believed so deeply in the work you were doing that she was willing to overcome her natural introversion and tell her story. Her authenticity helped the callers see that their hard work opened the doors for a lot of students who are hardworking, passionate, and motivated but cannot afford to go to school.

A lot of people are doing work that makes a difference, but they don't get to see for whom or how. Once you get to see that, you're much more willing to give your time, energy, and creativity to this cause that might have previously mattered to you but seemed too abstract to understand.

DP: *According to conventional wisdom, successful people have three things in common: motivation, ability, and opportunity. They need a combination of hard work, talent, and luck to succeed.*

Your research in Give and Take *focused on interactions with other people and team dynamics, which led you to identify three different reciprocity preferences: givers, takers, and matchers. I'm curious, why is reciprocity important, and how does it signify a fundamental shift in approach to management?*

AG: I think it's important because we all live and work in a connected world. If you track any industrialized economy, what you'll find is that the majority of people are working in service jobs, teams, or both. And literally, if you're working in a service job, then your entire goal is to help somebody else. If you're in a team, then your success is heavily dependent on how well you collaborate with and contribute to your colleagues. It's hard to succeed without having effective interactions and meaningful relationships—that is what drew me to the importance of reciprocity.

It's hard to succeed without having effective interactions and meaningful relationships

This is increasingly important in the sense that we've seen a dramatic growth of the service sector, so there are fewer people working in manufacturing and production kinds of jobs. The world of work has become more relational. There is also the rise of teamwork. People used to work independently on pieces of a product that would then get assembled, and now we're seeing more interdependence and back-and-forth collaboration, which makes interactions more important than they were before.

DP: *Reciprocity and connection are keys to understanding the workplace, which makes you think that maybe givers totally defy conventional wisdom and are always on top. Your research showed that givers were on both ends of the spectrum in terms of being the most productive and also the lowest performing. The correlation of giving to productivity and performance is contextual. How does giving relate specifically to success?*

AG: I think about givers as people who enjoy helping others and regularly do it with no strings attached. They made up both the most successful and the least successful performers across all types of jobs from engineering to medicine to sales.

The least successful givers may not be thoughtful about how they help, and so they end up helping takers who exploit them or they help in ways that don't energize them. This may lead to burn out or not contributing much. They drop everything to help other people, so they're busy doing other peoples' jobs or they run out of energy to do their own work, as opposed to blocking time to achieve their own goals, separate from their contributions to others.

The time you spend solving other people's problems makes you better at solving your organization's problems

Successful givers are thoughtful about whom, how, and when to help. Giving has all of these unexpected advantages to it. One is learning. The time you spend solving other people's problems makes you better at solving your organization's problems. When givers lend their knowledge to other people and go out of their way to share their expertise, they acquire new skills and diversify their capabilities.

The second benefit is social capital. In the long run, people want to work with givers. No one wants to trust a taker. Givers end up building broader and deeper networks, have more good will, and that social capital pays dividends when it comes to relationships and occupation.

A third benefit is motivation, which ties back to the fundraising callers we discussed earlier. Givers are able to tie their actions to something larger than themselves. They experience more meaning and purpose than their peers, which energizes them to work harder, smarter, longer, and more creatively. That tends to work out in pretty exciting ways and can drive success.

DP: We live in an era when massive changes in the structure of work and the technology that shapes it have further amplified the advantages of being a giver. For example, regardless of whether someone is a giver or a taker, service providers would hopefully be givers and on-demand workers would focus more on the team, shared goals, and these types of things. Can you help us understand why giving is so important now? Are we moving away from traditional hierarchies and reorganizing around teams?

When a giver helps, it feels like an expression of commitment and concern, like "I actually cared about this relationship and was making an investment in it."

AG: Work increasingly gets done through networks. Your ability to solve problems is partially a function of who you know, not just what you know. Givers have a much easier time reaching out to people in their networks because they constantly give help and established a track record. What I call "matchers" also believe in fairness or reciprocity, like, "You helped me, now I help you."

However, this isn't simply about activating connections with your network. Givers are known as good, decent human beings. People like to surround themselves with and gravitate themselves toward givers. What's interesting is that takers and matchers do plenty of helping too, but it creates a different impression that can feel transactional, like, "I didn't really care about you. I was just doing something for you so you would help me back." When a giver helps, it feels like an expression of commitment and concern, like, "I actually cared about this relationship and was making an investment in it."

The other reason giving is so important is that as the world's gotten more interconnected, you can't take advantage of somebody in one relationship and then start over fresh. There's usually an opportunity to trace back your common connections through LinkedIn or Facebook or other vehicles. Networks are not as invisible as they used to be. It's more likely that what goes around comes around. Givers have an easier time with their good reputation following them, and takers are easier to punish than before.

DP: This trend of interconnection and the visibility of networks that you describe will be more pronounced in the future, with ratings and reputation becoming key drivers of employment in the context of the gig economy and on-demand workers. The more data points and positive correlations to giving, the more likely someone is to get hired. This makes me wonder if givers are so important to successful organizations, how do you screen out the takers and turn takers into givers?

AG: Screening out is important because the negative impact of takers on culture is usually double or triple the impact of givers. The basic idea is if you can weed out the takers, the givers will be more generous because they don't have to worry about the consequences. The beauty of matchers is that they will follow the norm. With the presence of givers, matchers give too. You can shift a whole group by getting rid of the takers.

One way to do that is look at how people explain success. There's a common belief that takers use more "I's" and "me's" in everyday conversation. That's false. Takers use the same amount of "I's" and "me's" as the rest of us. You start to see disparity only when talking about collective achievements. Takers will take personal credit, where givers and matchers give credit where credit is due. Takers also blame their failures on others.

Dan Gilbert, founder of Quicken Loans, says the most important thing he looks for when hiring is: Does someone have a victim mentality? When you get into "What's a big stumbling block or failure in your career?" and then you ask, "Whose fault was it? What led up to it?" When somebody says, "Look. All these other people screwed me over," that could be a red flag.

You can often learn who is a taker by how they predict the behavior of others because people tend to project their own tendencies.

For example, ask a question about, "How common is it for other people to steal?" or "to hog credit?" Takers anticipate more selfish behavior than the rest of us, because that's how they justify and rationalize being a taker. They think, "It's not me. People are selfish, so I'm smart by being self-protective and cautious." If somebody anticipates that other people are takers, ask how they came to that judgment. The risky answer is, "I think at the end of the day, everybody is fundamentally selfish." What takers don't realize is they are giving a good look in the mirror.

The other question is, "How do you turn takers into givers?" I don't necessarily think that always plays out. We have lots of choices about these styles but they tend to be rooted in values, beliefs, and worldviews that are sometimes easy to change and other times pretty sticky. The best thing is to observe the fluctuations in somebody's behavior and look for moments where they're less selfish, more generous, and ask, "What do those moments have in common? Is there a certain role that they enjoy helping in? A kind of skill that they love to share?" Then tailor to reflect the patterns that you see.

Beyond that, make it clear that taking has negative consequences. Help them see their reputation. Most people don't want to be known as takers. Being perceived that way is a reason to change because it impacts their opportunity to succeed. If you measure people's contributions to others and not just their individual achievements, it's harder to get away with being a taker. For example, if I'm tracked on how well I mentor a direct report and how much knowledge I share, then it pays to be a giver. Performance evaluations and promotion systems should track not just what you achieve, but how you affect the success of people around you.

DP: I think the conventional view is that giving can take away too much energy or somehow make someone appear weak instead of being strong and assertive. You cite an example of a teacher that was at risk of burnout. After working long hours all week, she decided to volunteer on the weekends. Eventually, this decision helped her to become an award-winning teacher. It seems counterintuitive that spending more time giving instead of focusing on her personal health and resting on the weekends would prevent burnout. I'm curious, why is that?

AG: You have the punch line here already. First of all, people don't just sleep all weekend. They're going to do something. The question is: How are you going to spend that time?

A lot of people think that givers burn out when they're giving too much. At the extreme that's true. If you spend 100 hours a week helping other people and you don't take care of yourself, then that is a recipe for disaster. But there's a more compelling case to be made that givers burn out when they don't feel like they're having an impact and are not appreciated, or feel like their work doesn't matter.

In this particular case, we've got a teacher working in an inner-city school feeling like her teaching wasn't getting through. She was really struggling to help her students. Many of them didn't show up for class and weren't engaged in the learning experience. She volunteered on the weekends with a group of high-potential, underprivileged students. They were identified and selected on the basis of their future possibilities and were more invested in the learning process.

You need to feel like your contributions are making a real, lasting, meaningful difference

That rejuvenated her by showing, "Look. I am having an impact. I can have an impact." It restored confidence in her ability to give. To bring this full circle, I think this is part of what we did with the fundraisers. They're raising all this money and they don't know where it goes or if it's useful. Once they see it matters, they have these reserves of energy that they can invest to do the job. You need to feel like your contributions are making a real, lasting, meaningful difference.

DP: *In* Give and Take, *you champion this new paradigm for success that highlights the importance of giving, empathy, and altruistic behavior in the workplace. And in your latest book,* Originals: How Nonconformists Move the World, *you present a bold new vision for why people need to battle conformity, break tradition, and stand out. How has your thinking evolved, and can you tell us what led to your latest book?*

AG: One of the most frequent questions people ask is, "OK, I work in a culture of takers. How do I change it? Am I better off jumping ship for another organization?" Those questions are tied to research I've done for a few years on: How do you speak up effectively? How do you challenge the status quo? How do you champion change, especially if you're not the senior leader in an organization?

This led me to focus on originality. There are lots of ways to champion new ideas that can reduce the likelihood that other people squash them. That is different than trying to change your culture from taking to giving. There are many kinds of new ideas that people struggle to champion, where they feel like they are not fully understood or other people don't see why these new ideas are worthwhile.

My thinking evolved from "How do I as an individual in the world navigate my interactions?" (which is the focus of *Give and Take*), to, "If I'm in a system that doesn't make sense, whether it's a company or a community, how do I adjust the way that it operates and how do I effectively challenge what's wrong around me? How do I make sure that I'm not somebody who allows an undesirable status quo to persist?"

DP: *The marriage of giving and originality, or nonconformity, seems like it would create an environment that encourages the open exchange of ideas, focuses on more efficient serving of clients, resolves issues with client services, and embraces rapid innovation. In a sense, giving and originality become interdependent.*

AG: I would like to think that is true. I did a study with a colleague a few years ago showing that when you got people in a giving mindset, they actually became more creative. When people focused on producing a creative idea, they were seduced by the most novel ideas or the ones that were most interesting to pursue. Whereas when directed to think about how they can help others, that gave them a filter for separating the ideas that were novel, but not practical, from ones that were novel and could be beneficial to others. In this way, the giving mindset can make people more original.

I think to your point, it goes in the other direction too. As you think about championing changes, you have to figure out how they will help other people. If it's always about you, people won't want to make adjustments to the way that the world works. They have to see if there's something in it for them or for the group. In this way, championing original ideas could be an act of giving or force you into a mindset of giving.

As you think about championing changes, you have to figure out how they will help other people

DP: *When I think of originality and nonconformity, what comes to mind are breakthrough innovations by bold entrepreneurs. We're arguably living in a period of the most rapid innovation in human history, so there are lots of examples of nonconformists ruling the world, like Steve Jobs, Mark Zuckerberg, Elon Musk, and so on. How does nonconformity and originality translate into the workplace on a day-to-day basis for people working inside companies?*

AG: People underestimate their own originality. When they think of originals, they think of Steve Jobs, and they say, "That's not me." We all have ideas for improving our world, right?

Originality in the workplace translates through things like speaking up with an idea or suggestion to make a process work more efficiently. For example, we studied people at Google, and it was amazing how many opportunities Google salespeople had to think: "Here's a better way to serve our clients. We have all these clients who want ads, we're basically doing the job as it was handed to us."

I had a former student say, "You know what? I'm going to go visit a bunch of my clients," and nobody on his team had ever done that before. They did all of their communication by phone and email. It deepened the relationship in meaningful ways, and he found it to be one of the most worthwhile aspects of the job. That's an everyday active nonconformist. Everybody's managing the relationship one way; he thinks, "You know, maybe there's a better way to do this."

The funny thing is, it starts to catch on with the team and then you begin to wonder who is the conformist. We can all look at the organizations around us and say, "There must be ways that we can help them be more effective. We can challenge outdated traditions and practices that don't make sense."

DP: You mentioned Google. In a talk at Google, you jokingly said that you may have never become a professor if you knew at the start of your career that you could work for a company like Google—it institutionalized giving and collaboration, and is purpose driven in terms of not being evil. I found that interesting because I dropped out of my Ph.D. studies at Princeton to work in startups, also unaware when I started my academic career path that this type of work was possible.

You have coauthored a series of op-eds for The New York Times *with Sheryl Sandberg, COO of Facebook, sit on the board of Lean In, and are co-authoring a book with her. How are management practices and leadership styles coming out of Silicon Valley and companies like Google and Facebook impacting the broader workforce and the type of things that you teach your students at Wharton?*

AG: I think they're having a huge impact. There are so many cases where I've seen leaders in other industries say, "Well, you know, Google hires this way, so now we can give it a try," and "Hey, Facebook decided that work-life balance is really important, so if they can do this, we can too." It's exciting to see so many people thinking about a better way to organize work.

Google launched their re:Work website to share as many of their HR practices as possible, and that in and of itself is having a huge impact. Five years ago, a couple of colleagues and I started working at Google on people analytics research to help make HR and talent more data driven. At the time, to our knowledge, Google was the only company doing this. Fast forward and now we have 400 Heads of HR, Talent, and data scientists coming to a conference we host at Wharton, asking, "How can I implement better data-driven management? How can I bring more research into how I think about HR and talent?"

DP: We have the Googles and the Facebooks of the world leading this new style of management. Then in an op-ed for the New York Times *called "Friends at Work? Not So Much," you noted that in 1985, about half of Americans said they had a close friend at work. That number dropped to 30% by 2004.*

The research shows that friends are more trusting and committed to success of one another, and opportunities to form friendships make jobs more satisfying. Yet the numbers suggest this isn't happening. In the article, you attributed the decline and move towards a transactional view of work in part as related to Max Weber's thesis of the Protestant work ethic, which as a former scholar of religion I find interesting.

If there is all of this interest in collaboration and sharing, and tons of data to support why this behavior is good for us, why is work becoming more transactional and less friendly? What legacy issues do you think we're dealing with that need to be overcome?

AG: Lots of people think about work as a place where they go to be results-oriented, productive, and efficient. Especially as we move toward 24/7 connectivity, the pressures that people feel to make sure they're not falling behind has grown because they can be reached anytime and work never ends. You can also make a case that people are less invested in making new friends when it's easier than ever to stay in touch with their old friends through social media.

Part of the reason people are still drawn to the Googles and Facebooks of the world...is they foster a sense of community at work

Part of the reason people are still drawn to the Googles and Facebooks of the world, aside from the missions and work that they're doing, is they foster a sense of community at work. That's increasingly rare in a world of employment where nobody stays in the same job as long as they used to. Most people bounce from 10 to 12 careers over the course of a 30-year period. It's less likely that people are thinking about the place they currently work as the place they want to invest in building relationships.

DP: When I interviewed Seth Godin for my last book, he spoke about the idea that people need a unique voice. The press constantly talks about the next Mark Zuckerberg or the next Richard Branson, but that encourages you to conform to somebody else's ideal instead of nurturing what's unique and original. Seth also has this idea that innovators need to be separated so that they have autonomy to test and fail without the pressure and constraints of how the rest of the company operates.

How does the idea of originality and giving relate to innovation, and how should collaborative leaders strike a balance between encouraging innovations and making sure that their company is run efficiently?

AG: It's helpful to have separate innovation units, where the job that somebody has is to come up with and test new ideas. Maybe that is separate from the actual day-to-day work of running the organization. You can also bring in the rest of the workforce by running an innovation tournament, where you come out with a focused call, like "We're looking for solutions to this kind of problem and we're going to take as many ideas as people can submit. We're going to bring in subject matter experts to evaluate the proposals." The tournament might be a few weeks. This becomes a great source of variety and you get lots more people involved in idea generation. You also make sure that they're still doing their jobs.

DP: We covered a lot of ground around sharing and originality, how giving can be a driver of success, and innovation within the workplace and larger organizations. The collaborative economy is all about sharing and exchanging products and services on platforms created by entrepreneurs that rejected the status quo. In this sense, giving and originality, or nonconformity, are the drivers of the collaborative economy.

One of the main ideas of this book is that we need to focus on empowering each other with the tools and resources to succeed, and through mutual support and help we can co-create a better future. What would a fully realized vision of the world look like if everyone embraced giving and originality?

AG: Overall, people would be less paranoid. They wouldn't worry as much that other people are looking to one-up them, steal their ideas, or shoot down their suggestions. They would instead approach most of their relationships by assuming that other people were there to help. If they had an idea for how to make something better, they could speak up without feeling like they were going to get squashed. That would allow much more diversity of thought. People who have great ideas could come together more often as opposed to working separately in silence. A lot more innovation would emerge through these kinds of trusting collaborations.

Companies in turn might focus more on partnerships and think less in terms of cutthroat competition. Leaders could be more willing to take risks and tackle some of the big problems that we face as a society because they would be less concerned about failing and short-term results. We could reevaluate how we educate our children, train people for jobs, and think about innovation. In this way, the world would get incrementally better.

ADAM GRANT has been recognized as Wharton's top-rated teacher for five straight years, and as one of the world's 25 most influential management thinkers and *Fortune's* 40 under 40.

Adam is the author of two *New York Times* bestselling books translated into 35 languages. *Originals*, on how individuals champion new ideas and leaders fight groupthink, is a #1 national bestseller and one of Amazon's best books of February 2016. *Give and Take,* on why helping others drives our success, was named one of the best books of 2013 by Amazon, the *Financial Times,* and the *Wall Street Journal*—as well as one *Oprah*'s riveting reads and *Harvard Business Review's* ideas that shaped management.

Adam received his Ph.D. and M.S. from the University of Michigan in organizational psychology, finishing it in less than three years, and his B.A. from Harvard University, magna cum laude with highest honors, Phi Beta Kappa honors, and the John Harvard Scholarship for highest academic achievement.

PEERS INC.

Robin Chase

Robin Chase pioneered a new model of company creation when she co-founded Zipcar, and in doing so she changed consumer behavior forever. Paying for access instead of ownership, trusting customers to make transactions without an employee present, selling and tracking assets in real-time on a local community platform—these are pillars of what Robin calls Peers Inc. companies.

- Excess capacity and leverage drive exponential growth
- The balance of open vs. closed platforms
- Why self-driving cars alone get us nothing
- How connectivity can help address climate change

Peers are the community of users on platforms, and Inc. refers to the scale we associate with industrial capitalism. I included Robin in the opening section on Collaborative Leadership because her Peers Inc. model represents the future of how all companies will run and operate. She also presents a beautiful vision for how innovation can be a force for social good, generating value and improving the world.

DP: Peers Inc. *describes a new organizational paradigm that combines the benefits of small-scale localization and customization with the quality and scalability we associate with the industrial. The book also draws upon your experience as Co-Founder of Zipcar, a pioneer of this new type of company. What is the Peers Inc.*

model, and why are these new types of companies better than what existed before?

RC: A lot of people are seeing the rise of this new way of doing things, with many different names for it: the Platform Economy, the Gig Economy, the On-Demand Economy, the Sharing Economy. What I like about the Peers Inc. thesis is that it calls out the reality that this new economy is a collaboration involving two parties, each with distinct and important roles to play. The larger entity (the Inc.) builds the platform: simplifying complex processes, making use of economies of scale, applying standards. And then small, distributed peers (people, devices, data) deliver the local, customized, and specialized components.

Because they leverage excess capacity and invite the co-investment of peers, they can grow at exponential rates

We see this structure in companies like Uber, Airbnb, Kickstarter, Zipcar, Facebook, and Google, as well as in Bitcoin, Yelp, Massive Online Open Courses, and the impending 3D printing revolution. Even in governments such as Estonia moving citizenship and company creation onto digital platforms. These narrow and efficient structures are intended to harness and leverage excess capacity—a critical component that transforms the economics of these new companies—and invite the participation of diverse peers. Together, Peers and Inc. give us everything that we love about the industrial economy—things like scale, low prices, high-quality standards, and consistency—as well as everything we love about the local— unique, personal, adapted, creative, innovative.

Organizations built on this new collaboration grow faster, learn faster, and adapt faster. Because they leverage excess capacity and invite the co-investment of peers, they can grow at exponential rates. Think about the growth of Airbnb and Uber, or any app downloaded on a user's smartphone: The company didn't have to make the consumer buy an expensive phone to use their service. Instead, they are making use of the "excess capacity" available on the person's already-paid-for phone.

Peers Inc. organizations can learn exponentially faster because digital platforms enable serious analysis of gigantic amounts of variation and data. The platform is able to analyze and make use of all of the variation that peers deliver. Peers Inc. organizations can optimize their growth by determining and encouraging best practices, and spotting and discontinuing worst practices. This type of learning and pace was previously unimaginable.

Peers Inc. organizations also give us the chance to have things that are hyper-specialized. I think of it as "the right person will appear" or "the needle will fly out of the haystack" because the platform attracts such a diversity of offering and can reveal relevant details at the very right moment. It is because the peer collaborators are so diverse that these organizations can experiment, adapt, and evolve quickly at low cost.

Our new ability to connect easily and quickly has fundamentally changed the way businesses get the most value out of resources. Today, Peers Inc. collaborations extract the most value, innovation, and resilience through shared assets, shared networks, shared intelligence, and shared opportunities. It is a higher-value way to move forward than the old way that induced scarcity by heavily guarding a company's assets, employees, and intellectual property. Instead, it is the shared collaborative economy that creates abundance.

DP: *Zipcar was one of the pioneers of the Peers Inc. model. You processed the entire transaction through a platform—from purchase online, to hardware and software, to unlock cars and get keys—and lowered transaction costs down to almost zero, vs. $8–$12 for traditional car rental companies.*

Zipcar also tapped into the wasteful economics of current car consumption models, where personally owned cars tend to sit idle 95% of the time. This was a pioneering example of what became a pillar of the collaborative economy, which you refer to as unlocking excess capacity.

Can you expand upon that and walk us through the early days of Zipcar? What type of assumptions were you testing, and what obstacles did you overcome?

RC: Zipcar did a number of things right, and early.

We built a platform that gave our customers the powers that used to be reserved for employees only. Our customers choose specific cars and open that car themselves (with a proximity card). They could also see and manage their own accounts (this was back in 2000!).

We thought of our customers (the peers) as collaborators and asked them to do things that used to be done by company employees. For example, we ask our members to walk around the car and decide whether it is in good condition. We ask them to use our credit cards to refuel the tank when it is ¼ full. Before Zipcar, nobody imagined that rental companies could trust the customer. They thought an employee and customer had to go around the car together checking boxes that said, "A scratch on the left," "This is all right," "This fuel tank was full," and so on before and after the rental. It was inconceivable that there would be an element of trust and you would let the customer do all of that.

The difficulty of the technology was really the largest challenge and most amazing feat, but investors were most stuck on the change in behavior in car use. They didn't believe that we could rent cars without having a person there to do that walk through before and after, and they didn't believe that people would want to rent a car rather than own it. If we think about the whole peer-to-peer model and the sharing economy, Zipcar fundamentally proved that you could trust people, that you didn't have to have your own employee on the ground to make that trust handoff, and that there was demand for shared assets.

DP: *Zipcar changed behavior and also inspired new types of entrepreneurship. You not only pioneered the trust-based transactions, but the ease and accessibility of Zipcars at scale made it possible for people to not need to own a car. Zipcar in turn inspired other entrepreneurs to found similar companies, because if you don't need to own a car, maybe you don't need to own other things.*

In the book, you mention that your initial goal for Zipcar was "to make renting a car as easy and convenient as getting cash from an

ATM." We wanted to make clear that like ATMs, Zipcars would be available everywhere, 24/7, and take just a few seconds for the transaction. In hindsight, you credit success to making it easier and more convenient to rent a car than to own one.

This points to the concept of access vs. ownership. What is the significance of access models to the broader shifts in a Peers Inc. world?

RC: Start with the basic question: Why do we own things? A large part of the reason is to make sure that it's always going to be there when we need it. We used to think "I can't lend out my car or I can't have it part-time because it won't be there when I need it." In academic language, we call this a "rivalrous good." Zipcar showed that by creating pools of cars, this rivalrous asset suddenly became non-rivalrous. I could share it and still have it at the same time.

The music industry is an interesting example because it's gone the whole cycle—back in the olden times, you bought records or CDs because by owning them, the songs would be there whenever you wanted them. Then we unbundled songs so that you could buy just one song you wanted instead of the whole set. Now we've moved into Spotify and Pandora, where you don't need to own any songs because you can listen to whatever song you want on demand at any moment. The shift has happened because it is more convenient and cheaper to enjoy music this way than the old way, which required going to the store, buying physical versions of the songs, storing, managing, and caring for them.

In the case of Zipcar, I think people always would have liked to pay for a car just when they needed it instead of having to own, park, and maintain it. We demonstrated that this dream was indeed possible: "You know what? I can have a car only when I need it." Our tagline was "Wheels when you want them."

I worry about the "on demand" label. It makes people seem rude, privileged, and entitled. I snap my fingers and I get something. But I don't think that was really what was driving Zipcar customers or others in this new way of consuming. Rather, people are realizing that it is more convenient, more economical, and more . . . rational . . . to access something rather than own it full time.

It is more convenient, more economical, and more...rational...to access something rather than own it full time

DP: *That makes me wonder what other hidden assumptions we have about ownership and attachments to things that are breaking down. Access in some contexts has the same connotations that you associate with ownership, this entitlement and sense that this is mine. I wonder how we get away from this binary thinking of mine/not mine.*

RC: One of the key innovations with Zipcar was that each of our cars felt unique. Each car has a name and a particular location where it lives, and so it feels very much like mine. I see it and recognize it when I pass by and when I drive it. People become attached to that particular asset because it is parked in their neighborhood, and they are going to see it again and again, unlike shared bikes that are all alike. When things are plentiful and all alike, I think that causes a little bit less care in the usage. If I had a box of a thousand ballpoint pens, I lose one, whatever. But if I have my fancy, unique, with-my-own-name-on-it pen, I don't lose it, because it's one of a kind.

DP: *People say the same thing about renting apartments or rooms off of Airbnb. They take great care renting something unique from a person, cleaning up after themselves and treating it like it was their own home, vs. recklessly throwing towels everywhere and leaving the place a mess like they do in hotels. Even when hosts charge cleaning fees and you know that a professional cleaner is coming after you leave, there is still a level of care and consideration.*

That leads into another interesting question. BlaBlaCar, Airbnb, and Zipcar are examples of very closed platforms where peers can operate only within certain constrained and limited choices, resulting in relatively uniform collaboration. Then, on the other hand, you have these wide-open platforms such as the iOS and Android, open APIs and GPS. Can you help us navigate the difference between open and closed platforms?

RC: If you think about closed platforms, the platform creator has an idea about how that asset is going to be used. Typically, that means it will be used as it's always been used, just more efficiently. If we think about a closed platform like Zipcar, you're going to use the cars to drive. Airbnb, you're going to use places to sleep in because those are places to be slept in. Zipcar and Airbnb (cars and rooms) can exist in any geography and have varying price points, but the platforms are used in a specific way.

In open platforms, the end use or the means of the content use is wide open. For example, I can take photographs of my breakfast on Twitter or incite revolution by saying, "Let's meet up and have a demonstration." It's still constrained by 140 characters, but its use is wide open. The more open a platform, the more ideas people are going to put into it. That brings more innovation, so open platforms should ultimately have more value than the closed ones.

DP: Many readers of this book will be entrepreneurs or founders working on Peers Inc. companies. There is an infinite array of possible configurations between open vs. closed platforms. How do you figure out the right amount of structure and balance in thinking about organizing a platform?

RC: I think at the inception of their idea, founders often believe they need a lot more structure than they do. The objective early on should be to simplify! And apply a minimum of constraints. There's a kind of deconstruction that goes on in the very beginning to make your platform as open and easy as possible.

At the inception of their idea, founders often believe they need a lot more structure than they do

For example, think of Airbnb. Early on it shocked me that for years there was no way to sort by number of beds. You'd think the bed is the most basic thing. How many people does this offering sleep? That was not a field that they decided to include. Instead they minimized the number of fields for hosts to fill out. They settled on what was the smallest number of fields a host could fill out in order to deliver a viable space to rent. Airbnb sought to minimize the constraints to create a listing so that they could maximize supply, with the fewest possible barriers to creating a listing.

In the very early stage, you should have a conception of what you think your platform is going to be, and then work backwards and make it simpler. Add complexity later when everyone understands what the value is. You can't do that in the beginning. You should also choose a defined geography or sector in which you can deliver success to both sides of the market, buyers and sellers.

DP: You mentioned Twitter earlier, and in Peers Inc. *you talk about Twitter shutting down APIs as the ultimate betrayal for participating peers. Another example that you mention is AppGratis being shut down by Apple. Facebook similarly leapfrogged ahead of MySpace with an open platform that allowed apps developers to keep all of their revenue, and then years later it killed the apps ecosystem.*

This pattern seems to repeat itself over and over again. Successful platforms become monopolies and transition from open to closed systems. Then new companies come along to challenge the new monopoly incumbents. I'm curious, with so many billion-dollar unicorn companies in the collaborative economy space that appear to be carving out monopolies, how do you see the space continuing to evolve? Are we entering a consolidation or maturity phase? What's next?

RC: Right now, everyone's saying, "Is Uber really worth (whatever it's up to right now…like $70 billion or whatever crazy number)?" And I feel like it's way too early to tell. We're at the beginning of all of this.

If I had to evaluate valuations from a theoretical point of view, I would say that some of these are overvalued because of their monopoly status or the expectation that they will become monopolies.

Society only dislikes monopolies when they exercise their monopoly power—through predatory behavior that eliminates competition or high prices to consumers. But monopolies don't have to behave badly. The challenge is when monopolies are pushed, by CEOs or investors, to take advantage of their monopoly power. But remember that Peers Inc. companies have two parts: the platforms and the peers. If the platform gets too greedy, they risk angering the collaborating peers who may well choose to stop collaborating. Without them, there is no business.

Society only dislikes monopolies when they exercise their monopoly power

I feel like there's this hole in the periodic table, and I'm predicting that the hole will resolve itself. If I look and understand the structure of these peers and platforms, the peers should stop participating when they feel like they're getting screwed. We see drivers are on both Uber and Lyft, doing whichever one serves them best at that moment. New companies are coming along trying to offer a 5% take instead of a 30% take. Uber is also losing money to undercut competition, but that isn't sustainable. When the margins are better and new companies scale, what incentive is there for drivers to stay? I want to believe what Elinor Ostrom, winner of 2009 Nobel Prize in Economics, says—that the peers will stop participating if they can't make the rules of engagement and if it feels unfair—but that will take time to play out.

One of the problems with capitalism is that private sector companies are not playing for long-term gain; they're playing for a short-term gain. My prediction would be that companies that don't treat their peers well can't succeed in the long term. But sadly, right now in capitalism, we don't play the long game; we play the short game. So I don't know what will happen.

DP: You talked about the long-term game, and I would love to hear your views on the long-term game of transportation. Most of your experience comes from founding several different transportation companies. When I think of on-demand transportation services like Zipcar, Uber, Lyft, BlaBlaCar, etc., they're all becoming increasingly popular. Self-driving cars are also on the horizon, leading to speculation that car ownership may completely come to an end.

Self-driving cars open up all types of possibilities from allowing longer commutes because people could work or rest in their cars, to radical disruption to insurance industries and car manufacturing. We may even reimagine cities with less congestion and more pedestrian areas due to a decreased need for parking and other inefficiencies. Looking ahead, how do you see the future of transportation? What should we expect in the next 10 to 20 years?

RC: I expect autonomous cars will completely disrupt the entire sector. They will transform how we live in cities and how we spend our money. We will see the first mass market sales of AVs within 5 years (and as early as 2019), and I believe that within 5 years of that, the majority of vehicles in dense urban areas will be self-driving.

There's two ways that this can play out in terms of whether this is going to be a good thing or a bad thing from a transportation point of view. There's a high risk of what I think of as zero-occupancy vehicles. Right now, 80% of the cars on the road are single occupancy and people only think about the cost of driving as the cost of fuel, unless they're paying by the hour with Zipcar or by the minute with Uber, Lyft, or car2go.

Self-driving cars alone will get us nothing

Once you take human bodies out of the equation, it is cheaper to have that car drive around the block continuously than to pay for parking. It's also cheaper to send the car to Home Depot and back than it will be to pay for shipping. I think 50% of the cars on the road will be empty and that could lead to grotesque amounts of congestion. Someone commented in one of my talks that if you think of places like Mumbai, Cairo, and Delhi, those are effectively self-driving cars because the cost of labor is so incredibly low. Those cities have horrific traffic and pollution. Self-driving cars alone will get us nothing.

On the other hand, we can put a high penalty on zero-occupancy cars and encourage shared trips in shared cars. Add electric cars, and we get some cleanliness. What Zipcar showed is when you share the car itself, you reduce the number of cars needed, and that's critical. There's been a lot of research showing that we only need 30% of the cars currently in use. That will free up or eliminate on-street parking space. If we have shared trips in shared cars, then we address congestion and we may only need 10% of the cars.

If we have shared trips in shared cars (and address the zero-occupancy problem), we will transform cities. We can add more trees, more sidewalks, more bike lanes, and with no parking garages or parking lots there is a tremendous opportunity to do more with lots of developable space. It transforms access into opportunity. The Holy Grail would be a transit-priced trip door-to-door from home to work without the congestion. That is the potential if we get all of the policy pieces right.

But if we don't, I see a ton of missed cars delivering in vastly more congested cities. In addition to zero-occupancy private cars running around because they don't care, I think we'll see retail cars because having a car rolling down the street costs way less money than a retail store where I have to pay for the location. You can think of the liquor store, pharmacy, or the latte car roaming around, or pop-up roaming retail cars selling shoes and clothes or whatever. It all depends on what rules we put in place.

DP: Your latest company, Veniam, is building an Internet of Moving Things. The company proved out a model for mesh networking in Porto, Portugal, where Veniam transformed vehicles like buses and garbage trucks into mobile Wi-Fi hotpots, creating a seamless Wi-Fi network connection that allows people to go from one hotspot to another. Why is connectivity so important in a Peers Inc. world?

RC: If I think about all the things we've talked about, what is clear to me is that Internet connectivity is a mandatory part of this new economy. The ability to be networked-people and have access to networked assets is a fundamental requirement to a sustainable and just future. Without being accessible and findable, we can't get any of these benefits. The difference between the un-networked and the networked is going to become an even increasingly larger chasm than it is today, so we have to make sure that everyone is connected.

The ability to be networked-people and have access to networked assets is a fundamental requirement to a sustainable and just future

DP: You are a big advocate for addressing climate change, and at first glance connectivity doesn't seem like an obvious part of the solution. What type of unlocked potential is there on the horizon for the Internet of Moving Things, and how might connectivity help us to address some of humanity's most pressing problems like climate change?

Regarding climate change, smart homes and smart cities are all about a platform collecting and analyzing huge amounts of data to push forward best practices and knock out worst practices. Connectivity makes distributed energy easy to build and easier to connect it to a larger grid; we get the benefits of both the grid and the distributed networks. I see Veniam and this pervasive connectivity as a key element of all of that efficiency. Reductions in CO_2 and traffic congestion require shared, networked connectivity, with platforms collecting and analyzing the data, and empowering the diversity of the peers (people, cars, washing machines, renewable energy!) to deliver the energy efficient outcomes—all while delivering a high quality of life, of course!

....

ROBIN CHASE is a transportation entrepreneur. She is co-founder and former CEO of Zipcar, the largest carsharing company in the world; Buzzcar, a peer-to-peer carsharing service in France (now merged with Drivy); and GoLoco, an online ride-sharing community. She is also co-founder of Veniam, a vehicle communications company building the networking fabric for the Internet of Moving Things.

She is on the Boards of Veniam, the World Resources Institute, and Tucows. She also served on the board of the Massachusetts Department of Transportation, the National Advisory Council for Innovation & Entrepreneurship for the U.S. Department of Commerce, the Intelligent Transportations Systems Program Advisory Committee for the U.S. Department of Transportation, the OECD's International Transport Forum Advisory Board, the Massachusetts Governor's Transportation Transition Working Group, and Boston Mayor's Wireless Task Force.

Robin lectures widely, has been frequently featured in the major media, and has received many awards in the areas of innovation, design, and environment, including Time 100 Most Influential People, Fast Company Fast 50 Innovators, and BusinessWeek Top 10 Designers. Robin graduated from Wellesley College and MIT's Sloan School of Management, was a Harvard University Loeb Fellow, and received an honorary Doctorate of Design from the Illinois Institute of Technology.

#GIVEFIRST

Brad Feld

Brad Feld embodies the philosophy of giving first. Since 1987, he has played an astonishing number of leadership roles in the startup community—entrepreneur, investor, author, speaker, and co-founder of Techstars—helping to build great companies and sharing his wisdom via his blog and books like *Venture Deals* and *Do More Faster*. This interview is so jam packed with insight it's hard to summarize:

- Why giving first is part of conducting good business
- The Boulder Thesis to build a startup ecosystem
- The seemingly limitless potential of networks
- Why exponential change makes the future uncertain

Brad encourages founders to constantly learn about themselves and each other because building a great company is a journey that requires mutual honesty and support. He strikes a rare balance between strategic and technical thinking, and deep inner reflection, which makes this a thought-provoking read. *#GiveFirst* is also the name of Brad's next book. Here's your glimpse into what will surely be a bestseller.

DP: *You have been a long-time advocate of giving without expectation, or giving first. This philosophy is institutionalized in Techstars and your next book will be named #GiveFirst. What does it mean to give first, and why is giving first so important to the startup community?*

BF: The concept is pretty simple. The notion is that you are willing to engage in a relationship and put energy into something without knowing what you will get out of it. It's not altruism. You expect that you will get something out of it, but you don't know when, from whom, over what time period, or how. You enter into relationships non-transactionally, meaning you don't define the transaction or value exchange of things upfront.

The reason this is so important is that part of the challenge of startups—and startup communities in general—is that you have to get a lot of energy into the system. You have to get a lot of people working on stuff to make any progress. If everybody is trying to figure out what they are going to get out of it before they start putting in energy, then the startup community will be stillborn.

Giving first is mutually beneficial for everyone. It gets the energy and excitement flowing across a whole ecosystem so there is not this continual transactional tradeoff of expecting things in return. What happens is you end up getting things back from people or places that you don't expect because you were willing to put in that energy in advance without consideration of how you might directly benefit.

DP: *The conventional view is to think about giving in the context of philanthropy or nonprofit work, but in the startup community giving leads to innovation, deal flow, and new business. Giving breaks down barriers and gets people talking and helping each other, and that builds momentum and relationships of trust across the whole ecosystem.*

On the flip side of giving first, there is this notion that entrepreneurs and founders need to understand and know who they really are. The journey of being an entrepreneur is lonely and tough, and you can't waste time trying to imitate or be someone else. I spoke with Jerry Colonna about this for my last book Disruption Revolution, *and I know you are great friends that go back years.*

In an interview that you did with Jerry for his Reboot podcast, he recalls you giving him advice back in 1996. Jerry was struggling with what he referred to as "imposter syndrome" where he was stuck feeling inadequate and comparing himself to other VCs. You told him to "Stop trying to be a VC like everyone else. Be the best VC like you are."

That simple advice had a profound impact on him and the following year Jerry became recognized as one of the top 50 VCs in the country. This brings up the importance of radical self-inquiry. There is a way in which knowing the self allows one to understand his or her true purpose. How does that help make great entrepreneurs, and how do you see purpose coming together in founding teams?

BF: The first part is that we're on a journey through this thing called life, and at some point it ends and we die. It doesn't matter what you think happens after we die, but the particular journey construct that we have ends and is not static.

The most successful and interesting people constantly are looking for new things, trying new things, and learning new things. Some things work, some things don't, and they build on that. A big part of that process is internal and self-reflective—in other words, you're going on your own journey and you're learning about yourself, not just about the work you're doing. That leads to learning what you care about and want to do, how it works, and your relationship to what matters most to you.

This is a fundamentally systemic phenomenon for founding teams. You have to constantly learn about your cofounders, yourself, and your relationship to each other as the business goes from a couple of people and an idea to hopefully a very large organization. There is such incredible change that even if you built a large organization before (and by the way, even if you built a large organization with the same people you're building this new large organization with), the context, the backdrop, the stuff you're working on, the challenges you're going to face, the stressors, the exogenous world that impacts you in ways that you don't expect, etc.—they're totally different than the last time. You have to be on this continuous process of learning.

To link it back to what I said to Jerry about being the best version of himself, I think he had a lot of fear about being different that he understood as "imposter syndrome." Being different was part of his brilliance. He wasn't like every other board member or VC. That was why I liked working with him and probably why he liked working with me because I'm not like everyone else either.

This notion of a singular archetype—for a VC, founder, entrepreneur—is nonsense. Different people approach problem solving in different ways. You can learn from role models, but trying to emulate what you perceive to be an archetype of success will probably make you unhappy and may not lead to success. When you let yourself behave the way you want to behave, you actually work more effectively with people around you. That is why knowing yourself and your co-founders is so important.

> *When you let yourself behave the way you want to behave, you actually work more effectively with people around you*

DP: Boulder is one of my favorite places in the world. It has the highest per capita number of PhDs and people working in startups, ranks highest in quality of life, is home of University of Colorado, and with a population of around 100,000, Boulder feels intimate enough that you have a real sense of community.

Boulder is also the place that you call home, and your book, Startup Communities, *is based on what you call the "Boulder Thesis" for building a startup ecosystem: (1) Entrepreneurs must lead the startup community; (2) Leaders must have long-term commitment; (3) The startup community must be inclusive of anyone who wants to participate in it, and (4) The startup community must have continual activities that engage the entire entrepreneurial stack. Can you tell us more about The Boulder Thesis?*

BF: My hope is that today it's simplistic and fairly obvious. The phrase "startup communities" and the construct of The Boulder Thesis didn't exist in 2011–2012 when I wrote the book. At the time, we were dealing with a recession and many people were talking about a jobless recovery.

The book was a chance to look at the future opportunity for us as a society, in terms of economic growth, through new company creation and entrepreneurship. From my perspective, you have to take responsibility for your future. People who take responsibility for their future are the ones who have the most impact on society.

The way to do that in a noncommercial construct is to take responsibility for yourself, which is similar to what we just talked about. In a commercial construct, that means being an entrepreneur, creating your own business, getting involved in a startup community or entrepreneurial ventures, and helping them grow. This bottom-up approach with entrepreneurs as the leaders of the startup ecosystem goes against a number of top-down historical mythologies that I talk about in the book.

For example, there is the misplaced idea that the university is the center of the entrepreneurial ecosystem. Even if you look at cities like Boston or the Bay Area—the area around MIT in Cambridge, Boston; the area around Stanford in Silicon Valley—the university is an important input into the startup community, but it's not the center of it.

Another example is this mythology that governments should play a key role stimulating small business and economic development. How can government help? The answer is to allocate capital.

The problem is that governments have no capital to allocate. Coming out of the recession they had no money. They should have cut themselves in half and laid off a bunch of people, and then let them go try to do something productive. Governments are particularly bad at rationalizing their size—and people are constantly trying to justify their involvement.

This is not to say governments or universities are useless. Both can play a very interesting role. But the point of The Boulder Thesis was to shift the discussion to center on entrepreneurs. They need to be driving the activity proponents in the startup community.

Another key tenant of The Boulder Thesis is to be inclusive. This is not just in terms of diversity of gender, race, and ethnicity, but also diversity of thought, experience, training, and interests—to be inclusive of anyone who wants to engage. Today, there is this whole meta-layer of discussion about diversity. For example, think of what happened in North Carolina around sexual identification. Not just politically, but the economic impact in terms of companies like PayPal and Google Ventures not wanting to invest or conduct business there because they support diversity.

Be inclusive…not just in terms of diversity of gender, race, and ethnicity, but also diversity of thought, experience, training, and interests

DP: *It is important for entrepreneurs to be the leaders of the startup ecosystem, and also to be inclusive of everyone that wants to participate. Inclusivity has this expanded definition to encompass traditional conceptions of diversity, and diversity of ideas and perspectives.*

In your framework, entrepreneurs are at the center and everyone else serves as a feeder into the startup community. Can you explain the idea of feeders and the role they play in supporting the startup ecosystem? If the goal is to be inclusive, what is the best way to include non-entrepreneurs?

BF: First, there is one important clarifying point. It didn't occur to me in writing the book that some people would interpret "feeders" as a pejorative phrase or think leaders are more important than feeders. They're both important categories; they're just different in terms of what function they play.

Feeders are institutions or hierarchical organizations—universities, governments, large companies, nonprofits, etc. The powerful thing that I've learned over time is the best way for feeders to be helpful to the startup community is to engage proactively and have individual members of the institution engage as nodes in the startup community. Give responsibility to the individual person to become an active participant as a representative of the organization, instead of somehow having the organization try to manage or control the activity of the startup community.

This builds a positive reputation for the institution that you represent. Do something actionable instead of saying, "Hi. I'm from the Boulder Economic Development Council, and our job is to stimulate economic development." That doesn't mean anything. Get involved in the startup community and do something.

Give responsibility to the individual person to become an active participant as a representative of the organization

DP: Techstars was one of the first accelerator programs, founded around the same time as Y-Combinator. A key differentiator was that Techstars placed a greater emphasis on mentorship and the philosophy of giving first, which we discussed earlier.

Techstars has since evolved into a global ecosystem. There are Techstars accelerator programs across the U.S. and in many major international cities. This includes programs focused on particular areas, such as the Internet of Things, and accelerators powered by Techstars for global brands, such as Disney and Nike. What was the initial vision for Techstars and how has it evolved over the last decade?

BF: We were the second accelerator (YC was the first and we started about a year later), but I like to think we were the first to take this mentorship-driven model after which many accelerators have been modeled. We try hard to open-source our approach to encourage and enable more.

Our initial goal was an experiment in our backyard in Boulder: Let's see if we can start some companies and then see if it turns into anything interesting. That was the extent of the vision. We had some hypotheses, "If we can get a bunch of local entrepreneurs involved, it would be a good way to spend time with other entrepreneurs, working on a thing together that's not their own company."

We had ideas around the energy that would go into the local startup community—again, that wasn't the phrase for it then— getting a bunch of people for a limited period of time (not continuously, but for 90 days) working with a set of new founders. As experienced entrepreneurs and angel investors, we knew how satisfying it was to help other founders, which is hard to do in any quasi-organized way.

That was the starting point. We ran the experiment for the first two years. By year three, we started to have this notion of the power to build a significant network of founders, mentors, investors, and experienced entrepreneurs working together in support of founders who were creating new companies. There were lots of different ways to engage and participate that had quantitative economic opportunity as well as qualitative opportunity. As we had more companies go through Techstars, the network would get stronger and grow in an unmanaged or non-managed way, the same way the best networks grow, vs. the hierarchical "here's where you fit and here's what you do."

By around 2012, we started to view ourselves as having a significant opportunity around company creation and how innovation occurs. Building on the power of the network, we started to put more energy into ways to work with large companies to build startups and startup communities around their brands and their much larger orbits. It's a way to get much larger companies engaged in the innovation process. That's very powerful both in geography and across geography.

We're creating essentially a new type of innovation asset

I think that evolved another layer: We believe that we're creating essentially a new type of innovation asset. It's the opportunity for founders to engage in a much more effective form of company creation on a long list of variables or factors, but it's also a way for investors to engage in a broader set of companies that are highly curated in a way that previously weren't very organized or accessible.

If you're on the founder or entrepreneur side, this is a more effective way to quickly get your company started and then ramp it up. If you are looking to invest in early stage companies, this gives access to a category of early stage companies of high quality that's completely unique. Fundamentally underlying all of that is the notion that innovation (and the innovation activity that comes from entrepreneurship) is the most interesting characteristic around economic growth. Having this very large network that's constantly learning and evolving on how to do that is limitless in its scale potential.

DP: You mentioned the power of the Techstars network constantly learning and evolving, and how this has enabled the creation of a new kind of innovation asset.

The accelerator benefits entrepreneurs in terms of scaling great companies, and for investors it creates a pipeline of high quality, pre-qualified opportunities. The network continues to grow with every batch of accelerated companies and more accelerators, creating this seemingly limitless scale and potential.

I wonder if the power of networks can be utilized in innovative ways to create new types of companies. For example, imagine an Airbnb or Uber that shared ownership with providers or drivers, or a company built on the blockchain where shares and dividends were distributed based on the amount of work contributed by a decentralized group.

As a seasoned angel investor, venture capitalist, and co-founder of Techstars, what are your thoughts on new forms of company creation? Can you expand a bit on the power of the network and ways in which networks might be utilized in different ways to create companies?

BF: While I think this is one dimension, it has huge legal and regulatory challenges based on the current SEC rules around investment, solicitation, and company ownership. While there are some people trying to create new corporate structures that are disconnected from the current legal frameworks, especially in the context of blockchain, I think this will be difficult to do in any meaningful way.

But the concept can be extended to the notion of how a network of participants, separate from the legal structure, can be organized, especially around innovation and new company creation. If you take the notion of Kickstarter as a model for crowdfunding product development, you are using a network to build community, create market validation, and provide financing for product design and development. If you extend this to a non-employment model, you end up in what had a trendy, but short-lived label of the "gig economy." In each case, technology enables a network that enables a different operating model for an organization.

DP: There is this idea in the Boulder Thesis of a 20-year horizon for startup ecosystems, meaning that leaders should look 20 years ahead into the future from each day. We see lots of excitement on the horizon with self-driving cars, drones, AI, 3-D printing, the collaborative and sharing economy, virtual reality and augmented reality, etc. Yet with all of this innovation come concerns about massive loss of jobs due to automation. As a final question: How do you see the next 20 years unfolding?

BF: I have no idea. I'm not a predictor. I get endless interview requests to be parts of the lists of predictions for the future and I always decline to participate.

What I think is a truism is that human beings are particularly bad at understanding how things work on an exponential curve. We understand linear lines really well. When we extrapolate from where we're sitting out into the future on a linear basis, there is a very logical approach. But we suck when it's exponential. Much of what we're dealing with now in terms of innovation, technology, company creation, how humans and machines interact with each other, how our society is changing, etc., is on an exponential basis.

Which is to say that if you took a log of the graph over time, you'd get a line, and I just don't think we know how to think about it. One person predicts a 10-year horizon, and another predicts something will happen in 12 months. Or it could never happen because it's a completely wrong prediction. I really don't have any idea. I would assert that it's a particularly exciting time to be doing all of this stuff and to be in the middle because some vectors are changing at such an incredible rate.

We have these massive step-function changes in science and technology. For example, we're dealing with a bunch of stuff around new forms of computing, whether it be quantum computing or Biocomputing—a huge amount of research, lots of people struggling, lots of things interesting (maybe not working), and then you get this gigantic dislocation step-function where something completely magical gets understood and all of a sudden you've got a whole body of work that builds on this completely new framework for it.

Before you had to write books and physically ship them all around because there was no better way for people to communicate with each other. Now there is a bunch of software interacting with itself when all these things are happening. That's another example of how we don't know what happens with exponential change.

Twenty years from now, I think our planet and our society, and the way that we interact with it, will be unrecognizable to us today

Twenty years from now, I think our planet and our society, and the way that we interact with it, will be unrecognizable to us today. If we teleported forward 20 years, it would be unrecognizable. And I don't think it would be the same kind of unrecognizable as going backwards 20 years. If you went back to 1996 today, a whole bunch of things are very different in 2016 than 1996. But I think when we shift to 2036, it would be completely and totally unrecognizable to somebody from today.

....

BRAD FELD has been an early stage investor and entrepreneur since 1987. Prior to co-founding Foundry Group, he co-founded Mobius Venture Capital and, prior to that, founded Intensity Ventures. Brad is also a co-founder of Techstars.

In addition to his investing efforts, Brad has been active with several non-profit organizations and currently is chair of the National Center for Women & Information Technology, co-chair of Startup Colorado, and on the board of Path Forward. Brad is a speaker on the topics of venture capital investing and entrepreneurship and writes the blogs Feld Thoughts, Startup Revolution, and Ask the VC.

Brad holds Bachelor of Science and Master of Science degrees in Management Science from the Massachusetts Institute of Technology. Brad is also an art collector and long-distance runner. He has completed 23 marathons as part of his mission to finish a marathon in each of the 50 states.

COMMUNITIES AND MOVEMENTS

Douglas Atkin

Douglas Atkin is a pioneer in building communities and movements around brands, organizations, and startups, a trend that he anticipated in his book *The Culting of Brands* (2004). We draw upon his vast experience in the peer-to-peer economy as Chief Community Officer of Meetup, Partner at Purpose, Co-Founder and Board Chairman of Peers, and Global Head of Community for Airbnb:

- How small groups can change the world
- Why people join cults to feel like their true self
- How communities grow into movements
- Building the world's first super crowd brand

Douglas combines his passion to be a catalyst for meaningful change in the world with a relentless focus on research, often conducting hundreds of interviews and surveys to validate ideas and create actionable strategies. His leadership and example helped inspire me to follow my passion to write, speak, and build movements. I hope he also inspires you to live and work with passion and purpose.

DP: There's a quote by Margaret Mead that you absolutely love and reference in every presentation, and it is also one of my all-time favorites: "Never doubt that a small group of thoughtful, committed citizens can change the world. Indeed, it is the only thing that ever has."

When I think about the power of small groups, one thing that comes to mind is Meetup, where you served as Chief Community Officer. Could you tell us about the power of small groups and some of the lessons you learned building Meetup's global community?

DA: I love that quote because there comes a point where people realize that they need other people to pursue their passion or effect change. They can't be as effective on their own as they can be with a group of like-minded individuals. Whether it is a small thing like cleaning up a local park or a big thing like ousting a corrupt leader, you need help from others and the strength of mutual support.

If a community is strong, it's not just the power of like-minded people together, that's like 50% of what represents the community's power. The other 50% is about the interaction between those individuals. Relationships form when individuals interact. You want to support these people because you like them or they're important or they need help. Those feelings become mutual and form the bonds of real communities.

We discovered it took someone attending a physical Meetup event four times or more to become really committed. The reason why is because it takes that long to build meaningful relationships with other people in the group. You're going not just to practice your Spanish or your guitar or whatever, but you're also going to be able to see Pablo again, or Kia, or Jane.

When we asked people "Why do you like this Meetup?" they would answer by saying things like, "I feel at home," and "I made new friends." The original intention was to go and improve a skill or find others who share your enthusiasm. But once relationships form, then a community becomes real.

It's very hard to break a community apart once it gets to that point because people need and want to help each other. The relationships are strong, and the ties are strong between members. There is a power in a group that changes and transforms you to some degree in many positive ways. To bring this full circle to the Margaret

Meade quote, it can empower you to change the world, if that's what you're looking to do.

There is a power in a group that changes and transforms you

DP: We met years ago in New York City when I reached out after reading your book The Culting of Brands, *which looked at the cult-like followings of brands such as Apple, Ben & Jerry's, and Harley Davidson. Research for that book laid the foundation for your work in Meetup and Airbnb, and building movements that reach millions of people.*

The Culting of Brands *explored two questions: Why people have this cult-like commitment or loyalty, and what do you have to do to get that level of commitment or engagement? Can you tell us some of the lessons learned from your research that might apply to creating any type of community or community-related brand?*

DA: The reason I looked at cults and cult-like organizations is simply because those are the extreme forms of community. It is easier to understand the fundamentals of belonging and belief by looking at the extreme end of the spectrum. I also learned in the process that cults aren't an aberration. In fact, they are normal and essential because new ideas help keep cultures iterating, growing, and moving. Without cults, cultures would atrophy and die.

The key lesson from cults that can be applied to all communities is what I call the great cult paradox: People join cults not to conform but to become more individual. Most people think the opposite is true—that people join because they're psychologically flawed or socially inept. This is due to the media's portrayal of cults that are destructive organizations. Most members of cults and cult-like organizations join for the same reasons that you or I would join anything.

It works like this. As we grow up and become individuals, we realize that to get on in the world you have to shave the rough edges off of you, your identity, just to get on at school, not to be bullied, to form groups of friends, get on with your family, at work, etc. This doesn't mean that your individuality completely disappears, but rather that you compromise in a way to fit in, unless you can find a group of people who share the same differences you have.

Basically, what cults say is, "Hey, Douglas, you're different. We're different in the same way. Come and join us." That difference could be anything, such as a passion for big motorcycles. Harley riders used to tell me this. Despite having a fantastic job, being a management consultant or dentist by day, lovely family, a good suburban house, that wasn't them. They only felt at home when they were among like-minded others who shared a passion for rebellion and freedom, which they felt Harley users did.

People join cults not to conform but to become more individual

When you find other people who share your passion and, like you, feel different in the same way, you feel "at home." You psychologically relax and feel secure that no one's going to laugh or ridicule the things you're passionate about. In fact, they love you for your differences. That creates—I kept hearing this again and again—a psychologically safe space to become yourself. They use the word "become."

People joining cults and cult-like organizations—even the Marines or a corporate cult—all said the same thing: It doesn't change you; it enables you to be more yourself. And this is because you feel "safe enough" to express your unedited self.

DP: You mention the Marines and corporate cults. How does the loyalty, sense of belonging and purpose, identity formation through differences, etc., relate to building community in the workplace?

DA: Most recently, I've been working on the internal community at Airbnb. I found the same story here. For example, one woman expressed it by saying, "I can be my full fat version of myself rather than the skim milk version."

DP: That's a great quote.

DA: Yes it is! People told me that when you don't feel at home in your job, you edit yourself because you don't feel secure or safe enough to express yourself fully. It's normally only with your closest friends or members of your family that you can do that, unless you're lucky enough to find a company, or a church, or Meetup group, or whatever it is that makes you say, "A-ha! I've finally found where I feel at home." What that basically means is I feel psychologically safe and secure enough to be myself like I normally am only at home.

Brian Chesky, our co-founder and CEO, wrote an essay that was widely shared on social media titled "Don't Fuck Up the Culture" on the advice Airbnb's investors gave him. Culture is so important to maintain as you grow.

Airbnb culture is incredibly strong. Not everyone would like it, but those who do, really do. They say the same things and use the same vocabulary: "I can become myself." "I can be myself." "I don't have to edit." "I feel psychologically safe and secure because I'm surrounded by people who welcome me for who I am and celebrate my differences." "They don't criticize or laugh at me."

Google also studied what makes teams successful. They identified five characteristics, the most important of which is "psychological safety," as they call it, which is the same thing I'm talking about. What they found is that no matter what teams are working on, they need to feel a kind of security that only appears when you trust each other. And you trust each other because you know and respect each other.

It happens because that team is a community. You have relationships built through individual interactions. You know each other's strengths and weaknesses and trust each other. Relationships of trust and mutual support create the psychological safety to be yourself. That enables you to take the risks you need in order to learn something new, to be entrepreneurial, highly creative, or take any kind of business or personal risk. This all goes back to what we talked about earlier from my experience building communities at Meetup. The same thing happens in communities everywhere, including the workplace.

Relationships of trust and mutual support create the psychological safety to be yourself

You need to be surrounded by people who are creating a psychologically safe work environment. New team members arrive as strangers. They're welcomed, made to feel they belong, and then they're given some huge, crazy challenge. Some describe it as "I feel I'm supported by my teammates and my boss to take a massive risk." Other people might not think I can succeed, but then they do and realize, "Oh, I can do it after all." This would never be possible without that psychological safety net of freedom, trust, and support from their community of colleagues.

DP: I'd like to shift from communities to movements. You describe movements as basically like communities in action, or communities on the move. After The Culting of Brands, *you left advertising to be a partner at Meetup, and you co-founded Purpose, a consultancy for movements that also incubated and created movements. For example, you helped build All Out, the largest gay rights movement in the world, from 2,000 to 2,000,000 people.*

Through your work at Meetup and Purpose, you developed a systematic approach to grassroots organizing and movements. I want to dive into some of the specifics around your approach. People always talk about the mission and purpose, but one of the things that I find especially interesting is this idea that it is important to have an improbable goal, which you refer to as the "fuck off" goal or "fuck off" metric. I love this term. Can you tell us what you mean?

DA: The purpose of social movements is to make a difference and change in the world. They do this by mobilizing huge numbers of people to take the same action (such as signing a petition). One of the most important ingredients of their success is having a seemingly impossible goal. It isn't actually impossible; it's just improbable.

Let me give you an example, marriage equality. I'm gay. I'm married to my partner of 26 years, but even five years ago we thought the goal of the equality movement seemed impossible. It was never going to happen, with a Republican congress and so on. But it has. This is the difference between impossible and improbable.

Many things that seem impossible are not impossible, just improbable. The reason why you need the "fuck off" goal and ideally a "fuck off" metric is that you need to create a vision that's worth all of the effort. It has to be literally visionary, as in *I see a new world*. Like Martin Luther King said, "I have a dream." Basically, you're saying I have a vision of the world that doesn't exist yet but should. That's exactly what your vision needs to be for an organization, whether it's a company or a movement that wants to make a positive dent in the world. It needs to be desirable enough to say, "Yes. I will do all these hard things to help make that become a reality."

Basically, you're saying I have a vision of the world that doesn't exist yet but should

Ideally, you also have a "fuck off" metric. For example, with All Out we wanted equality everywhere; for gay, transgender, lesbian, etc. people, and we have a metric for that. When we launched, there were 76 countries in the world where it's illegal to be gay and 10 where you can be executed or receive life imprisonment for being gay or transgender. Our fuck-off metric was to go from 76-10 to 0-0.

DP: This same idea of a big "fuck off" improbable goal can be applied to the vision for companies.

DA: Yes, exactly. For example, at Airbnb we take extremely serious our vision of a world where "Anyone Can Belong Anywhere." We know we can make that world happen in millions of small ways through the actions of our hosts. They enable strangers to feel like locals by welcoming them and weaving them into the social fabric of their neighborhoods and cities. We're developing a metric that will enable us to measure how much guests felt they belonged. It will be used to measure our success, together with our business metrics.

DP: In contrast to the big "fuck off" goal that's improbable, you also have this idea of the commitment curve, where grassroots organizers start with a basic first action that is a minimum level of commitment, such as signing a petition or joining an email list, and then move on a curve through stages of increased participation up to attending a rally or physical event.

Can you tell us about how a commitment curve works? Why is it important to start at the bottom instead of jumping ahead to the big ask?

DA: The commitment curve is a very simple model that enables you to make "asks" of your members and users such that a larger number of people become committed, and committed more completely, to your movement or organization. The top axis measures the degree of commitment. The bottom axis is time. The commitment curve travels from lower left to upper right. The idea is that you plot asks on the curve from easy at the bottom, to hard at the top. Start with a low-barrier-to-entry ask, such as signing an online petition.

Next, follow up with a slightly harder ask. Not a massively larger ask, but a slightly harder ask. For example, tweet at your Senator or post something on Facebook. It takes a little bit more effort and a little bit more personal commitment, but you do it. Then you might follow up with a slightly harder ask, such as making a donation.

The idea is to make increasing, incrementally harder asks, which in turn lead to incrementally more commitment. Any given ask on that commitment curve is only slightly harder than the one before, so it never seems like a huge jump. Then before you know it, you find you will have ramped large numbers of people up the curve to ever-higher levels of commitment.

DP: At Airbnb, on one hand, you have these decentralized communities of hosts around the world that are self-organized. There are forums that hosts can join and provide mutual support to each other. They can also form meet ups and take innovative initiatives. One example you mentioned in a talk was a woman started a peer-to-peer group where people could sell items and raise money to decorate places and become hosts.

On the other side, Airbnb proactively engages communities to take action around legislation or to engage in their communities to raise awareness for the economic benefits of home sharing. As the Global Head of Community, can you tell us about the balance between these decentralized, self-organizing communities vs. the more hands-on, proactive engagement of communities to take action in the form of grassroots organization.

DA: We created the community platform because we know that hosting can be hard, and the mutual support of other hosts will make it easier and more fun. For example, a new host may be thinking, "Oh my God, what do I do? Do I have to put clean towels out; do I do this? So I do that?" Getting tips and advice directly from other, more experienced hosts can be enormously helpful.

Conversations are about everything from "How do I create a good welcome for a honeymoon couple?" all the way down to "Oh damn, my washing machine's broken and I need to change the bedding. Is there someone nearby whose washing machine I could use?" The basic idea is that hosts help hosts become better hosts. And that's exactly what happened.

It's a new economy bumping up against old laws in cities around the world

Grassroots organizing is a different methodology that's required for a different purpose. Airbnb hosts and guests are participating in the peer-to-peer economy. It's a new economy bumping up against old laws in cities around the world.

We've used grass-roots organizing techniques to invite our hosts and guests to become part of the political process in their cities, and to create laws that both recognize this new economy and are fair to their fellow citizens. Grassroots organizing is truly excellent at scaling the effects of community.

Unlike the traditional community manager model that exists in most organizations (usually startups), grassroots organizers identify and then train members of your community to become community leaders. In others words, they try and make themselves redundant by recruiting and training your members to be leaders.

In San Francisco, for example, we had 11 host-leaders that ran communities of hosts in each of the 11 Supervisor Districts (Supervisors are the governing body in the city). The organizers trained and equipped them to tell their stories to their representatives effectively, to attend hearings and testify, to give press conferences and so on. They used the commitment curve to make the right asks at the right time. The result was a new law that involved unheard of community involvement in its making.

DP: You have such a wide range of organizing experience, from building small groups at Meetup to global communities and movements through Purpose and Airbnb. Most people today talk about "community" in generic terms, as if all communities are the same regardless of size or interest.

Can you help us understand the different nuances based on your extensive experience?

DA: Let's go back to the first two things that we talked about, movements and smaller communities. Never be confused by thinking they are the same thing; they're not. They exist for different purposes and offer different benefits.

A smaller community like a Meetup group, or a PTA or church, exist to create a sense of belonging, so that people can learn, do, or change something together. Social ties between members tend to be strong within small intimate communities.

It's different for movements. The whole point of movements is not to be small and intimate; it's to have massive scale. Your goal is to mobilize large numbers of people to take action on a single leverage point (normally a government, and ideally a person, such as a Prime Minister, a Mayor or whatever). The social ties in a modern movement tend to be weak. But the effectiveness comes from many people taking the same action at the same time.

DP: I want to ask one final question and build upon this idea of world firsts. Airbnb is arguably the biggest crowd brand in the world doing pioneering work at massive global scale. Recently, you launched the Belo, a new logo that is also a symbol for people, places, love, and the letter A.

I've heard you speak about the Belo in the context of launching a global super brand, in the sense of being a universally recognizable icon similar to a Coca Cola or Nike. There is this big idea behind the Airbnb brand that anybody can belong anywhere, and you encouraged your 1 million hosts and 25 million guests to embrace the Belo and make it their own brand. People created and uploaded over 80 thousand versions of the logo to your website. Airbnb is basically going into uncharted territory as the worlds first super global crowd brand.

As Global Head of Community and as a former branding expert, this seems like an awesome culmination and triumph of your life's work. Can you tell us about the Belo and the story behind launching the world's first super global crowd brand?

DA: I first joined Airbnb as a consultant. I thought it was to help them with community because I'm a community guy. But Brian said, "Hey Douglas, you know a lot about brands from your past, can you help us figure out what ours is?"

"We are clearly a community, I can see that," I said. "There are three stakeholders in the community (hosts, guests, and employees.) Employees are both hosts and guests; and guests are also hosts; and vice versa. We have this massive community. I think the question to ask is: What's the purpose of our community? Why does it exist? What is it in the world to do and what difference is it going to make? In other words, what is the vision and how will this community make the world a better place?"

Once we have established what that is, then we'll know what the brand should become. But the brand is only one manifestation of the vision. Vision will also be manifested in product design, our office space, and who we hire—all of those things.

We ended up talking to 485 hosts, guests, and employees. The stories that people told, and the data from surveys, suggested that the community's purpose was something in the area of, "We are trying to create a world where anyone can belong everywhere." Eventually it became "Belong Anywhere." That's what was used to brief the Design Studio in London that developed the Belo with our in-house design team.

When we launched the Belo, we wanted to launch a symbol, not a logo. There's a big difference between the two. A logo is simply a graphic design. A symbol is a graphic design that has a meaning attached. We simultaneously launched the Belo (our symbol) with the idea of Anyone Belonging Anywhere.

Brian, Joe, and I flew to New York to show both the symbol and its meaning to some guests and hosts. We shared the story of "Belong Anywhere"—what it means, and where it came from. One host from Brooklyn said, "Thank you for explaining to me what it is exactly that I do and why I do it." That's when we knew we had something that really resonated. Some people even had tears in their eyes!

When we launched the Belo and "Anyone Can Belong Anywhere," we also invited our community to submit their own versions of the Belo to our site. Over 80,000 people submitted their own versions. This just reinforced the idea that the Airbnb brand is a community brand.

One of our current marketing campaigns called "Live There" is another great articulation of the brand and our vision of "Anyone Belonging Anywhere." It says don't go and visit somewhere, go and live there, even if it's for a night because that's what it feels like. Feel like you're so immersed in the culture and have an inside track on the place that you feel like you're living there, even for a short time.

DOUGLAS ATKIN is Global Head of Community at Airbnb; Co-Founder and Board Chair of Peers.org, a Global Movement for the Sharing Economy; Founder of theglueproject, a blog and venture about social glue; and Board Member of AllOut.org, the world's largest LGBT movement. Previously, he was Co-Founder of Purpose, an organization that mobilizes millions for social change, and Partner and Chief Community Officer at Meetup—the world's largest network of communities.

Douglas is also a former brand strategist and partner at leading NY and London agencies and the author of *The Culting of Brands: How to Turn Customers into True Believers,* a book about how to build cult-like community around almost anything. He lives in San Francisco with his partner, Matthew, and two beagles.

PART II

THE POWER OF SHARING

How can we build a collaborative society?

Why does sharing make life more meaningful?

How can you provide and share value with others?

SHARING COMMUNITIES

Neal Gorenflo

Neal Gorenflo has one of the most inspiring personal stories you will ever hear. His radical life change from corporate executive to co-founding Shareable exemplifies the type of transformation that is possible when we let go of material definitions of success and pursue a life filled with meaning and purpose. For Neal, sharing isn't a thing we do; it defines who we are and and helps us realize our true potential.

- How sharing transforms lives and communities
- The power of open networks to make change
- Liberating work from the cage of jobs and money
- The importance of the commons to Sharing Cities

Neal has participated in thousands of sharing events and is a pioneer of the sharing economy movement. He has a dynamic ability to encourage participation from everyone in a decentralized, open network; and then shift to strategic thinking about policy recommendations on how to scale Sharing Cities around the world. What an inspiration . . . get ready to quit your job and start sharing!

DP: *What is Shareable and why did you found it?*

NG: Shareable is a non-profit with a mission to empower everyone to share. We have two programs: publishing and community organizing.

For publishing, we are best known for our online magazine at Shareable.net where we serve about 100,000 unique visitors per month. Our editorial strategy is to inspire and empower. The dominant narrative in society is that you succeed through shopping and competition, that it's every person for themselves. We tell another story—how to succeed through sharing and collaboration. We're all in it together. And we show through many examples in nearly every dimension of life that sharing works.

For community organizing, we bring together the people who put these principles into action. That's everything from how to host a potluck to starting a worker cooperative. For us, sharing is very broad. It's not just about sharing items but also sharing power, ideas, and ownership, so it extends from neighborly sharing where you borrow a cup of sugar to sharing a stake and a voice in an enterprise or government. Our community organizing connects people who share this vision and want to live it out and make it real.

About two years ago, we started the Sharing Cities Network. The basic goal is to foster strong, local sharing movements in cities around the world. The Sharing Cities Network connects sharers in a knowledge exchange at a global level. We are also working on a book to solidify a vision of what it means to create a Sharing City.

DP: *Many readers of this book want to share more, but might feel stuck in a job they don't like or they just don't know where to start. We tend to get trapped into this conventional way of thinking that sharing is something we do part time, instead of being an essential part of who we are, what we do, and how we interact with the world.*

I would love to start this interview with your personal story, because it is so powerful and inspiring. You had an epiphany that led you to leave the corporate world. Can you tell us a bit about that experience and the realizations you had on how to live a more meaningful life?

NG: June 2004, it was a sunny Saturday afternoon and I was staying at the NH Hotel in Brussels. I had been travelling a lot working for a big multinational corporation, DHL, part of the backbone of the global economy. I ate a hearty European breakfast and headed out on a jog through my normal route in this business park. When I got to the first turn at the top of the hill, I stopped in a parking lot of a warehouse.

Then something unexpected happened—I started to cry. I realized this was not what I wanted to do with my life and that I would never realize my creative potential on this corporate path. I would not have the type of relationships I want, become the type of person I want to be, do meaningful work, or be part of a real community.

> *It's every man for himself, every woman for herself. This is not the way to get what we really want out of life*

My life flashed in front of my eyes. I saw all of the personal struggles and challenges I had up to that point, and there was a thread that connected them all: I had been doing all life on my own and was deeply alone and lonely. Along with this, I felt hopelessness, disempowerment, and even shame that I hadn't been able to make a real life.

Even though on paper I was a success in terms of my career, I felt like a failure in life. I realized that this was the general condition many people feel in the capitalist system and corporate world. It's every man for himself, every woman for herself. This is not the way to get what we really want out of life, which is a basic desire to become a fully realized human being. This wasn't just about me, but also feeling that pain in others.

I made a decision on the spot to change my life and do whatever I could to create a world where it was easy to find love and friendship, community and meaningful work, and where every day could be filled with authentic human connection. Except, I didn't know how to do that. I kept asking myself, "What should I do?" and I didn't have an answer. I made a vow to find out. I ran back to my hotel room, sent in my letter of resignation, and booked the first flight home to start a new life.

DP: Wow, that's a powerful story. The moment at the end reminds me of the story of the Buddha, where after attaining enlightenment he basically stood there for three days wondering what to do next. Even with total clarity of mind on the inside, we sometimes don't know what to do or how to navigate the world.

I agree 100% that many people want to break out of the corporate world but don't know what to do. When I read through some of the policy docs, initiatives, and writing at Shareable, *I see a lot of different applications of strategic thinking from the corporate world. This makes me curious how you shift directions in life without starting completely over.*

How has your background been helpful in the work that you do now, and how might readers from similar corporate backgrounds start sharing initiatives or contribute more to their communities?

NG: It was helpful to have some experience, but I would say that the most powerful thing going for me is my level of commitment. It is deep and real, and it goes to the marrow of my bone and the core of my soul. I had this unexpected a-ha moment that I didn't ask for. I listened to it and acted on it.

My advice for people who may not be satisfied with their life direction or career is to provoke that kind of moment—to search and question, not necessarily to recreate the same thing as me, but to tap into their deepest desires and beliefs. That's a process and it takes a while. I had this moment, but there was a lot of preparation and build up to it. It didn't come out of nowhere.

Ask questions, be curious, listen and be open, particularly to what comes through you. One of my favorite quotes is by Martha Graham, the famous American modern dancer and one of the greatest artists in the 20th century:

> There is a vitality, a life force, an energy, a quickening that is translated through you into action, and because there is only one of you in all of time, this expression is unique. And if you block it, it will never exist through any other medium, and be lost. The world will not have it.

It is not your business to determine how good it is, nor how valuable, nor how it compares with other expressions. It is your business to keep it yours, clearly and directly, to keep the channel open.

That's what I would say to anyone in a similar situation as I was. The change can't come from just your intellect. The mind can help, but there has to be a spiritual shift, and that takes time.

DP: I love this. It's so important to remember that life is a journey and not a chain of decisions from job to job, project to project, venture to venture. The power of sharing comes from within and opens us to new possibilities in every aspect of our lives.

I heard you talk about a similar idea before. If you help someone with the thing that they are dying to do or that's the most important thing to them in the world, then they will want to help you back. By totally focusing on the intent and the purpose of helping others, you create this kind of network or ripple effect of profound change in the world.

NG: After I left my corporate career, I came back to San Francisco and started a salon with friends called the Abundance League and that was precisely the design. One evening each month, everyone connected based on their purpose: not their job and the thing that put money on the table, but the thing they were dying to do in the world, the unique contribution they could make. We connected on that basis to help each other on our passion projects.

If you help someone out with something that they are dying to do in the world, you make a fast friend who is willing to go the distance for you

If you help someone out with something that they are dying to do in the world, you make a fast friend who is willing to go the distance for you. If you do that hundreds of times, then what comes back to you is more substantial and meaningful than you can plan. That changed my life more than anything. The salon was the smartest, most meaningful, and most powerful thing that I've ever done.

It showed me how you can get things done in a different way that is organic, powerful, and without struggle. I am such a strong believer in sharing, generosity, and collaboration because I experienced the transformation personally. So have many of the people at Shareable. It is not just an idea to talk about. Sharing is our experience, and it's life changing, so we want others to experience it too.

DP: You mentioned Shareable, and how your personal commitment to sharing led to creating this salon and an organization to promote and facilitate sharing. Since you left your corporate job back in 2004, you have now led or been a catalyst for thousands of different sharing initiatives all over the world.

How do you find a balance between someone being a community leader vs. more open collaboration in a group? Or to rephrase that question: How do you balance a core strategy vs. letting things evolve organically over time?

NG: I think this comes down to the type of organization that you have, whether it is a traditional organization or an open network. Operating in an open network is very different from the way you would operate within a traditional organization.

You are not commanding people to do something. Instead, you make the invitation to join in what you are doing

There are different principles at work in an open network like the power of example, the power of suggestion, and the power of empowerment. You can get massive levels of impact by that approach. You are not commanding people to do something. Instead, you make the invitation to join in what you are doing and it's the power of your purpose that makes the invitation attractive.

The open network approach brings people on the stage with you to make change. This is how we operate at Shareable. It's partially a strategy, but also we adapt to things as we learn and see opportunities unfold. If you are at the center of a conversation and help start a movement, then you are in a unique position because you have a sense for what's going to happen next and you can prepare the way.

DP: This search for a more meaningful life rooted in sharing and collaboration led you to leave the corporate world and rethink things on a massive scale. Part of that involves the nature of work and the application of startup thinking to re-engineering how society functions.

There is a way in which peer-to-peer platforms and the sharing economy can evolve to tackle bigger societal problems and inefficiencies at massive scale through offering co-ownership to providers and members. Next generation companies that offer more value through co-ownership incentives could undermine the monopolies of first generation sharing economy platforms that charge fees per transaction.

Could you tell us your thoughts on the future of work and the direction that you see the sharing economy and peer-to-peer platforms evolving in the future?

NG: I think we see a limited and unambitious exploration publicly around the future of work that centers on how we can make this freedom-limiting institution better or make the cage more comfortable. This seems completely backwards and un-American in the sense that it's anti-freedom. It shows how conditioned we are in this economy that we cannot think beyond a job.

In a sense, this is like going backwards to before the 19th century. The labor movement had some focus on improving wages and working conditions, but the priority was to shorten the workweek over time until there was no work. In other words, the goal was freedom through prosperity and abundance. The real question we should ask about the future of work is: How do you get from a job to no job?

Let's try to imagine it from our contemporary situation. We have lots of on-demand workers on peer-to-peer platforms and things like InstaCart delivering groceries, etc. One possible transformation that is partially under way is these platforms could give their stakeholders (the users and providers) a say and an ownership stake. This would help platforms remain competitive. This kind of shift would allow you to go from just being a worker on a platform to also being an owner and a decision-maker, and perhaps have a say and stake in multiple platforms.

DP: This seems like a natural evolution of the sharing economy business model. There is this kind of assumption in Silicon Valley around companies that have incredibly high, multi-billion dollar valuations that they will maintain their monopolies. However, they often operate at massive losses, and it's unclear how they will ever make enough money to please their investors without doing something drastic that undermines their communities.

Meanwhile, more nimble startups that did not sell the majority stake in their company to venture capitalists keep pushing the sharing economy business model further by offering a share of ownership to their users and providers. You speak about this elsewhere with the idea that ownership is the new sharing.

NG: Yes, exactly. Imagine you are an owner and decision-maker in platforms, instead of working for wages. Perhaps you occasionally work to get some income but maybe you are getting dividends or building up equity to help these platforms be successful. That stake in ownership also provides additional incentives for you to work harder, be loyal, refer friends and family, and so on. Ownership could create more sustainable business models instead of this hyper growth fueled by venture capital that is ultimately unsustainable.

That's a vision for platforms. But you can have a similar arrangement for housing, food, and transportation using less technology-intensive modes like cooperatives. Instead of buying what you need by working and earning wages, you are a member of a community that creates, manages, and uses a shared asset—i.e., a commons—that provides you benefits like food, water, electricity, transportation, housing, etc. That is where we need to go.

Then the question is: What do you do with all that? You start to get out of the trap of always having to work. The idea is that you want free time to develop yourself, which in turn develops into capacity to contribute even more to the community.

You have support for what you want to gift to the world, but you are creating a gift that only you can give to your community. That's a reality that already exists in part, but how can we create a world where everyone has a shot at that reality?

You start to get out of the trap of always having to work

DP: *This big picture, visionary perspective on how the future could or should be creates a great transition into your work on Sharing Cities. To start, could you explain what is a Sharing City and what types of initiatives are you doing at Shareable to empower them?*

NG: First, the idea of a Sharing City hasn't been defined and is in some ways contested. Corporations have grabbed hold of it and are defining it in a certain way, and nonprofit organizations like Shareable have a different perspective.

Our view is that a Sharing City is defined primarily as a commons rather than a marketplace or a political or governmental entity. That doesn't mean there isn't a market; it's that the commons is dominant. Examples of commons are things like public parks and cooperatives, co-working and hacker spaces, time banks and tool libraries.

A complete Sharing City doesn't exist in any one place, but parts of it exist everywhere. The effort we engage in is to bring all those pieces together into a singular vision. Every function of the city can be operated as a commons, from utilities and transportation, to food and housing—people working together as peers in a commons can manage most if not all of a city.

DP: *One of the great things about the commons is that everyone is equal, so it puts politics, career, job roles, class, and economic issues aside for common interests. You mention that parts of Sharing Cities exist everywhere, but there isn't a complete Sharing City.*

What advice would you give to people that want to participate in the commons and live up to the ideals and principles of a Sharing City?

NG: You want to build a movement. First, find others who believe in what you believe in and create community around the idea. Next, find shared priorities. Then come up with common projects.

I don't think it should be done like a bureaucracy where there's centralized decision-making and a single plan or strategy. Rather, empower people to work on the things they are passionate about and are able to execute. Don't get hung up on a singular focus because that can dissipate the energy and be a roadblock to action.

Have a portfolio of projects that work towards a common vision, even if that is lots of small things that you can get done. In Silicon Valley, there's a mindset that only values high-impact, scalable ideas. That leaves so many people off the stage of action. It is also elitist and un-democratic. It's not that we don't need those kind of big solutions or that we shouldn't be strategic. Rather, the fetish for the "big idea" at the exclusion of everything else wastes a lot of potential.

The fetish for the "big idea" at the exclusion of everything else wastes a lot of potential

Start with small projects like neighborhood potlucks, clothing swaps, and little free libraries. They get the ball rolling and are powerful cultural artifacts that signify something important about the community—that everyone here is in it together. The new relationships that form around these small projects lead to other often larger projects, which is the more important and almost always overlooked thing. Spillover effects like increased social capital can shift the trajectory of an entire community toward ever increasing levels of sharing and cooperation.

DP: *This gets back to what you spoke about earlier in terms of an open network. People come together with shared intentions and a common purpose, and projects spark ideas and become catalysts to think bigger, draw in new members, and grow in unanticipated ways. There is still this common core identity of the community, but it's not hierarchical. Can you give us an example of how a small project created this kind of snowball effect that transformed the entire community?*

NG: One of my favorite examples is the Sungmisan neighborhood in Seoul, South Korea. The catalyzing event was a community-lead campaign to stop the development of a water treatment plant that would have necessitated the destruction of their neighborhood forest. This campaign put in motion a chain of events that resulted in a 700 family urban village.

First, they created a childcare co-op. That led to a consumer co-op and then an entire elementary school, which was unique. It was a public school, but they incorporated peer learning. The elders teach kids things like carpentry, pottery, photography, and so forth. Then they formed a collection of clubs for hiking, farming, parenting, and photography, and that led to annual festivals and theatrical productions.

This village went from a protest campaign to an entire village economy where everyone is deeply interconnected. There's a sense that they are all in it together. That's why these little neighborhood projects are so important. One small thing can be the start of a community heading in a totally different, positive direction.

One small thing can be the start of a community heading in a totally different, positive direction

DP: That is a great example of how an entire community changed through sharing. I know that you cover these types of things on Shareable.net, but it makes me wonder why we don't hear more stories like this. It reminds me of the work that I did years ago when I ran the Religion and Conflict Resolution Program for the Tanenbaum Center for Interreligious Understanding.

We studied the efforts of religiously motivated peacemakers that put their lives at risk to resolve conflict through nonviolence. Part of our challenge was that nonviolent conflict resolution doesn't make headlines. Media would cover almost any story about extremism and violence, but a community of people reconciling differences and working together wasn't deemed newsworthy. Your work and that story of the small village in Korea are important reminders of sharing and collaboration all over the world.

NG: Yes, this kind of thing happens all the time and no one talks about it. They don't get the headlines because it isn't easy to talk about like one platform or one CEO. I believe we're entering an era where the community is becoming the hero. They are the central change makers. It's people coming together voluntarily, not because anyone tells them to do it, but because they choose to do it.

This is heroic. This is what we should admire in society. Collective actions like this lead to the type of answers that I sought when I left my job in search of a more meaningful and connected life. We can think big about changing the world through Sharing Cities, and we can start locally with small projects in our neighborhood. It is all interconnected. That is the power of sharing.

....

NEAL GORENFLO is the co-founder and Executive Director of Shareable, a nonprofit that publishes the world's leading online magazine about sharing and supports the global sharing movement through campaigns. He is a speaker, consultant, and writer on sharing cities, the sharing economy, travel, and the future of work. He is the co-editor of the sharing-themed books *Share or Die, Policies for Shareable Cities, and How to Share.* As a pioneer of the sharing cities movement, he advises mayors, communities, and organizations around the world how to meet their goals through sharing.

....

LOVE IS THE ANSWER

Prince EA

Prince EA has a unique style of inspirational content that embodies the types of values we need to co-create the future. He claims that he has nothing to teach; he merely helps reveal what is found within us all. His humility and authenticity helped him organically build a platform of around 6 million fans and a billion video views—a feat even more impressive considering his medium is spoken word poetry.

- Why the only true reality is in the present
- Mindfulness combats the roar of social media
- The ideal way to collaborate with influencers
- How self-awareness could shape the future

Prince EA covers topics like our addiction to social media, how labels of race and gender create divisions, and our impact upon the environment, yet his videos come back to a simple message: Love Is the Answer. A 28-year old native of St. Louis, Prince EA has become a global ambassador of peace and an international phenomenon. I'm grateful to share his wisdom with you.

DP: I heard that your name Prince EA comes from a 6,000-year-old story about the Prince of the Earth. It seems like such a great fit for the vision that you have today to grow the world through motivational, inspirational content. This makes me wonder if you always had a vision of transcending the label of "hip-hop artist." Can you tell us about the origin of your name?

PE: Prince EA, Prince Earth, comes from a place called Sumer. I was totally fascinated by this culture—they had so many technological advances right out of the Stone Age like political systems and inventing the wheel. They created language, cuneiform, letters with the characters etched in the stone. How did they know so much?

If you asked a member of this society, they would have said that their living gods bestowed this knowledge upon them. Prince EA, Prince Earth, Enki was the creator god of the Sumerian people, and he freed them from bondage through knowledge and wisdom. The name was so fascinating to me. I took it upon as my moniker because through my music, through my art, through my film, I want to free people—at least, to open their minds through knowledge and wisdom.

Before that, my name was Richie Rich. My real name is Richard. I was rapping about money I didn't have, cars I never drove, women I didn't get with. I was always very lyrical, and I rapped about what I saw, but it was very braggadocio. I wanted to be more authentic.

DP: *Authenticity—that is definitely the first impression I had when I came across your videos. There are so many people trying to be 'inspirational' but your content seems to tap into a deeper source. My sense is that the evolution of your public persona and content was parallel to a more personal transformation. Can you tell us about your personal journey and how that shaped what you create?*

PE: I come from St. Louis. The FBI just came out with statistics that called St. Louis the most dangerous city in the U.S. I grew up with drug dealers and criminals all around me, though I was lucky enough to have my family—my mother and my father—here. I knew that I didn't want to waste this opportunity that I have in this mystery of existence by following in the footsteps of another man. I wanted to carve a new path and my desire for truth, understanding, and happiness led me to teachers like Papaji, Mooji, Thich Nhat Hanh, and Ramana Maharshi.

I think the West advanced technologically and they look outward, and the East advanced technology of the heart and they look inward.

The external world is created from the internal. Shakespeare said, "Nothing is good or bad, but thinking makes it so." Is reality different from how you perceive it? Realizing that allows you to relax and know that everybody's doing the best that they can.

We're not here for very long; we're just walking each other home. I want to squeeze as much juice out of this experience called life as I can. We live a very ephemeral life on this planet. The worst thing we can do is to die without having lived and so I wanted to live a full life. I wanted to create and be my truest self, be vulnerable in my music, in my spoken word poetry, and so that's what I did. That's how the evolution happened. I began to create pieces of content that were meaningful out of service.

We live a very ephemeral life on this planet. The worst thing we can do is to die without having lived

My content reflects my inward journey. I take experiences that I've gone through, look inward to my reactions to things, and watch my emotions. Joseph Campbell said, "If you want to help the world, what you must teach is how to live in it." That's what I do. I teach people how to live in it, because there's one Earth, but there are billions of worlds. We're all trying to figure out our individual world, so I bring it back to the individual and created this platform. This is me. This is the story.

DP: I love this idea of one Earth, but billions of worlds. It reminds me of a powerful line in one of your recent videos "Everybody Dies, But Not Everybody Lives" where you apologize and say, "Martin Luther King didn't have a dream; that dream had him. People don't have dreams; dreams choose them. And we all need the courage to live out the dreams we have inside."

Many of the people reading this book are entrepreneurs and innovators who are constantly filled with doubts and struggle with the courage to pursue their dreams. Even when they succeed, at times they experience imposter syndrome, where they feel like an imposter for looking successful, when deep down they are consumed with doubt. Do you remember when your dream chose you? How did you overcome doubt and have the courage to pursue your dream?

PE: One word: death. Perhaps the greatest teacher of how to live is death. There's a book called *The Tibetan Book on Living and Dying*. In the beginning, it quotes the Buddha saying, "Don't think. But if you must think, let it be on the certainty of your death." The Buddha says that of all footprints, those of the elephant are greatest, and of all meditations, that upon death is the greatest. Death reveals what's important. It takes away the ego and leads you to realize the ephemeral nature of life and existence. Every moment is so fleeting, this unrepeatable miracle that we call "moments."

We have to live our most authentic selves. The path to serenity is that which is not reliant upon a certain result. We must give up the desire for a certain result and put all of our effort into the action itself, the creation, and the service. The result will be up to God, the universe, Buddha, Allah, I don't know. But all we can do is our best. When we realize that we don't know if we have 50 years, 5 years, or 5 minutes left, then it dissolves the ego. We've got to be here now and find out how can we live the truest versions of ourselves today. Not tomorrow. Today.

DP: What you just said touches on this idea of awakening this power within us all. You have this great quote on your Facebook page where you say, "Prince EA is devoted to growing the world through motivational and inspirational content. He has nothing to teach; he merely shows people the power that they already have within themselves. He believes that Love is the answer to every problem we face on this planet, and will continue to spread that message as far as he can." Can you tell us a little bit about the role that you see yourself playing, in terms of helping to awaken or show this power within us all?

PE: I have no grand ideas about some universal or global raise in consciousness or awakening. I don't know about any of that stuff. All I try to do is create content that's important and that can reach as many people as possible. I try to make my message very relatable. I use humor, and what the Buddhists call "skillful means." I try to speak the language of society and the generations of today to convey these messages because, as you said, I really have nothing to teach.

I think a lot of it is simply unlearning. Once we begin to question our thoughts and beliefs, we step into a space of vigilance and awareness. That state of awareness is our true being. Every belief, every label, every nationality, it is not original to us. I simply want to show people who and what they are and do that however it may unfold.

> *A lot of it is simply unlearning. Once we begin to question our thoughts and beliefs, we step into a space of vigilance and awareness*

DP: You organically built a following of millions of fans and your content has reached hundreds of millions of people. I wanted to interview you for this book because I think you're a great example of how social media can have a positive social impact and purpose.

Yet in some of your videos, like "Can We Auto-Correct Humanity?", you talk about how social media keeps us from interacting with others and building meaningful relationships. We've become glued to our smartphone screens and obsessed with status updates, forgetting about the people around us.

How do we find balance, and what lessons can we learn from your example in terms of using technology for social good without letting it control us?

PE: You said it: "Without letting it control us." We must maintain our position and not allow it to dictate our lives and our thoughts. The piece "Can We Auto-Correct Humanity?" wasn't about throwing your cellphone in the Mississippi River or getting a sledgehammer and breaking your new Samsung Android phone. It was about mindfulness. If you had a very romantic dinner with your wife or loved one, then you don't need a cell phone to distract you. Your mind could be on the future, on the past, a million miles away from the present moment.

Once we can develop a state of mindfulness, a maturity of mindfulness, it doesn't matter what we do. Anything can distract us. The cell phone is probably the most modern-day distraction that we have—the Internet and social media—but we can be present once we cultivate complete mindfulness, which is the best place we can be. It's the only place we ever are really. We just don't realize it. Once we're here, fully present, that's when we can truly create a great future and reconcile the past.

To get back to your other point, I think social media and the Internet allows our minds to roar and it has certainly done that with me. It also allowed me to spread my messages far and wide, from Portugal to the Democratic Republic of Congo. It's very beautiful to connect each other, all of us together, and I love it. But like you said, we cannot allow it to control us and dictate our thoughts and behaviors. We must maintain a state of awareness and mindfulness and I think everything will be all right.

DP: I want to shift directions a little bit and talk about your creative process. Your most popular videos have reached 50 to 100+ million views and are several minutes of spoken word, which in itself is amazing, considering the average attention span on social media. You tackle some complex subjects like the environment, labels, and depression, and how technology and social media can control our lives.

In an interview with Glenn Beck, you mentioned that for one piece you went through a kind of thought experiment about death. If you were going to die in a week, what would your last message be? From that place of vulnerability, there's only love and compassion. How do you strike a balance between being serious and giving depth to your subject matter vs. being light enough to reach a general audience?

PE: I'm a simple man. I don't consider myself to be extraordinarily intelligent by any means. I enjoy simplicity, and ironically, I think the highest form of sophistication is simplicity. Think of Einstein, his theory of relativity is an inch long and yet it explains so much. The profundity of it is universal and simple. When we frame things simply, that's how we get the messages out. That's what I try to do.

I don't consider my topics very heavy. I believe that everything is perfect as it is right now. Everything is perfect. My desire to want to help the world and to change the world is also perfect. When you can come from that place of "everything is fine as it is," it's a different energy, as opposed to "I need to change this. The world needs to be like this and this." There's still urgency, but it's just a different energy, and I try to come from that place.

When you can come from that place of "everything is fine as it is," it's a different energy, as opposed to "I need to change this"

Alan Watts has this great talk on perfection. He says to imagine if we had the ability to choose our dreams. The first would be a dream where everything is lovely and goes right, and the next dream will be just like that. We would get to a point where we get tired of everything going right. Then, we would choose a dream where some things go right, and some things go wrong. Let it be a mystery. Surprise me. And that is the reality that we exist in today. This mystery, this perfection, it couldn't have been better.

DP: This goes back to what we talked about earlier in terms of authenticity and teaching people how to live. There is this magical gift of life that we are all blessed with having, and we are all on this journey together going through life doing the best that we can. There are these kind of universal experiences to tap into, and yet each piece of content tells a unique story. Can you tell us a bit about your creative process?

PE: As far as my creative process goes, ultimately, the messages are the same. For example, "Dear Future Generations: Sorry," the message on the surface is about deforestation and environmental destruction, but when you dig deeper it's about knowing who and what we are, finding our connection with each other and the environment, to realize that there's no separateness. In the video "I Am NOT Black, You are NOT White" on the surface it seems like it's about cultural tolerance and racial harmony, but ultimately it's about knowing who and what we are deep down.

Osho talks about the problems in the world and likens them to leaves on a tree. When we pull one leaf from a tree, another one will grow. When we focus on the problems of society, racism, deforestation, all of these issues, another leaf will grow, another problem will come. We must get to the root of the problem. I like to get to the root and dig deeper.

When you realize that the real cause of these problems is identity and who and what we think we are, then imagine if we realized that we were all one. What happens to racism or sexism? There wouldn't be any legs left to stand on because that realization gets to the root of the problem. This isn't a belief that I want to try to convince people. When you look deeply, you know it, or you are it. My content is about taking you on a journey. This is the process of how I create.

DP: Social media has given rise to the category of influencers, these people whose recommendations and opinions influence millions of people to take action. Influencer marketing has also become incredibly popular. Most people would consider you to be an influencer, and I know you've done content and forged partnerships with Stanford Trees and Neste, for example.

Lots of people reading this book will be marketers or work with companies that are interested in partnering with people like yourself. What do you think is the ideal way to approach an influencer partnership so they have a positive impact that's mutually beneficial for everyone involved?

PE: The brand identities have to fit like Legos. I think that that's crucial. I've been contacted by a lot of different companies brands that were diametrically opposed to what I create and what I stand for.

DP: Because those people just see the numbers on your videos, fans, and that's what they want.

PE: Exactly. They should do their research on the influencers, have conversations with them to make sure that everything lines up, and then allow the influencers to have freedom to create. For me, my personal experiences have been good and bad. Sometimes corporations want to direct you to do or say things in a way that it's just not going to create the best outcome for anybody.

One piece of advice is to trust that the influencer knows how to message whatever it is that the brands are marketing. Trust that the influencers know what they are doing. Give them that trust and freedom, but to approach it collaboratively would be ideal. Make sure things align and trust the influencer to create—those are the two principles to stick by that will yield great results.

Trust that the influencers know what they are doing. Give them that trust and freedom

DP: I agree 100% that collaborative approaches are always ideal. You're doing a very cool collaborative project right now with Neste, which sounds like a great example. This book will come out when that project is still going on and unfolding. Can you tell us a little bit more about that?

PE: I was speaking in London at a sustainable brands event and met a representative of Neste, which is the world's leading provider of renewable diesel. They power the city of San Francisco transportation and are doing great things like creating fuel from waste. They are in total alignment with what I believe in terms of creating a sustainable society. We chatted a few times, and they flew me out to meet the CEO. They are a beautiful company trying to do great things for the world.

That led to a collaborative project called Pre-Order the Future. Essentially, we're giving people the opportunity to create a product. It can be in the field of education, transportation, entertainment, so we are getting all types of great ideas submitted to our website. These ideas are voted upon and at the end of the year, we will take the best idea and build that product out of renewable materials.

This project not only has the potential to create a successful product globally but also be successful in the sense that it will be in alignment with environmental integrity and help create a better, healthier world. It's something I'm very excited to create. We're in the process now of narrowing down the ideas, so by the end of the year we should have something great. It should be big. I'm excited.

DP: That's awesome! It sounds like a perfect example of the collaboration between a company, influencer, and the crowd, which is what this book is all about.

I have one last question: Years ago, I went to a talk by Thich Nhat Hanh that had a profound impact on me. There was a point where he said that the future can be like a prison. We have so many thoughts and expectations about how it could or should be that we become trapped instead of allowing it to unfold naturally. That talk was a spark that led me to quit my job, buy a one-way ticket to Thailand, and placed me on a path to write my first book, Red Bull to Buddha.

I mention this because the idea of being trapped by expectations reminds me of one of your most popular videos, "Dear Future Generations." The narrative opens with you apologizing for how we ruined the environment. There are no longer trees and our mindset was to call all of this destruction and pursuit of money and profits "progress." There's this dramatic buildup about how terrible the future will be, and then the music stops and you refuse to accept the future that we appear trapped into expecting.

That video has been viewed over a hundred million times, a great testament to the vision and inspiration of your content. So many of your videos look ahead to the hope of a better future, which is a main theme of this book. What would a fully realized vision of the future look like if all your ideas were implemented?

PE: A world in which we know who and what we are. Where we are no longer identified with who we think we are, with what we think about the world, what we can be. The Buddha was asked that very same question, "What do you think the future will be? Will there ever be an enlightened society? What do you think?" And he said, "Why are you worried about the future when you can't even stay in the present?" Nobody ever gets to the future.

There is no tomorrow. Yesterday is a graveyard. All of these are notions. The only true reality that we ever live is the now, and if we can stay here, present, my goodness. The future will unfold as it may. As it may.

There is no tomorrow. Yesterday is a graveyard. All of these are notions. The only true reality that we ever live is the now

PRINCE EA has touched the hearts, minds, and souls of millions of people worldwide. By producing creative, inspirational, and thought-provoking content, he has accumulated over 500 million views on the Facebook and YouTube platforms alone.

Born and raised on the North Side of St. Louis Missouri, the 28-year old poet, filmmaker, and speaker graduated Magna Cum Laude from a full scholarship at the University of Missouri St. Louis, with his BA in Anthropology.

Today, when he's not creating, he speaks at conferences and gives lectures to high school/university students nationwide, on the topics of self-development, living your passion, and the importance of being motivated and engaged in the classroom.

COLLABORATIVE SOCIETY

Antonin Léonard

Antonin Léonard is Co-Founder of OuiShare, a global community working to build a collaborative society around values of fairness, openness, and trust. Through empowering people and decentralizing power, we can find new ways to share value and create prosperity for all. This strongly aligns with the main theme of *Empower* to approach the future like we are building a movement.

- Decentralized organizations without hierarchies
- Why the collaborative economy isn't sustainable
- Changes coming from the next generation of leaders
- Grassroots movements, companies, and governments

For Antonin and OuiShare, many companies considered to be part of the collaborative economy remain trapped by an outdated model of capitalism. Massive growth accelerated by venture capital is unsustainable and creates risks of being forced to serve investors at the expense of the crowd. OuiShare is there to incubate the next generation of companies and ideas that put people first.

DP: The OuiShare website says, "Welcome to the Collaborative Society. OuiShare is a global community that connects people, organizations and ideas around fairness, openness and trust." You also wrote a book The Collaborative Society: The End of Hierarchies. *What is your vision for a collaborative society?*

AL: OuiShare started covering the collaborative economy in 2007 and we were among the first to approach it like a movement with different dimensions. At that time, "the sharing economy" and "collaborative consumption" were the two expressions mainly used by entrepreneurs, strategists, and thinkers covering those topics. To us, it appeared to be going beyond sharing and consumption, so we started using the term "the collaborative economy" as its main expression.

Today in France, the term "collaborative economy" is mainly used to describe Airbnb and Uber. Since day one, it was our intention to go beyond just covering those companies and show that there were a lot of movements converging together that were about empowering people and decentralizing power, basically giving tools to people for them to be entrepreneurs and producers instead of only consumers.

We eventually left the term "collaborative economy" in favor of "collaborative society"

That's why we eventually left the term "collaborative economy" in favor of "collaborative society." It's not only about economics or business. We are sharing a human-driven vision of economics in society. Right now, we are focusing on emerging initiatives—whether they are startups, nonprofits, or grassroots movements—that are empowering people. For example, we just did a five-week-long accelerator for open-source hardware and open-source manufacturing initiatives. We are slowly but surely moving from the startups in the collaborative economy space towards openness as a driver for innovation.

This is especially because these new models are struggling to be sustainable, whereas companies with massive amounts of venture funding like Airbnb, Uber, and others are not struggling at all. Pretty much the opposite. We are also in contact with policymakers in France and across Europe to help them address the economic and social impacts of these movements—but always with the vision focused on what it means for people to have a better life, be more autonomous, and have more power.

DP: Ending hierarchies and decentralizing power does not mean that you are anti-capitalist. Rather, you are pioneering and supporting the next wave of collaborative organizations with alternative business models and corporate structures, such as sharing ownership and co-investing from members, crowdsourcing, and cooperatives.

OuiShare applies these same principles in its own operations. You have local groups, knowledge groups, events and conferences, and also an online magazine. Can you tell us how OuiShare is organized, your membership and initiatives, and how these relate to your values and purpose?

AL: We are a very decentralized organization. There are a lot of initiatives, a lot of autonomy, and a lot of power to local chapters. For example, the team in Barcelona is doing their own Spanish-language OuiShare Fest, and we are trying to help them the best way we can. Sometimes we see them do things that we might do differently. We may have a conversation about it, but they decide what is best for them. This is because certain things might make sense for us in Paris, but for political or economic reasons, they don't make sense for Barcelona. We try to eat our own dog food.

Our leadership and governance are decentralized locally and globally. Every time there's a big decision that impacts the whole organization, we use an open source collaborative decision-making online tool called Loomio. The people that can use the tool are called Connectors and there are currently 80 around the world. Connectors are responsible for leading the community at a local level. Every six months, we all meet during what we call a summit. That's when we make decisions that impact the organization financially or in terms of exposure.

DP: OuiShare is a nonprofit with volunteers all over the world, and you have a major focus on collaboration and community. To what extent do you think the collaborative economy is a movement, and how do you empower it with this highly flexible, decentralized structure?

AL: First of all, I think the collaborative economy is too broad and it is not clear what it means. We stopped using that term. It's already old-fashioned, in a way, and it's too business oriented. We are creating and growing an organization focused on meaning, purpose, and having a positive impact in the world. Helping grow the collaborative economy might not make the world a better place, because the organizations that are leading the collaborative economy right now are very traditional.

What I mean is they are traditional in the way they raise money, how they work with people, and in their main focus on business. Many of them don't seem to be interested in something other than money and making their investors richer. I'm not trying to be anti-capitalist saying that. Our interest in the collaborative economy as a movement—and what we now call collaborative society—is based on an opportunity to create and grow different kinds of organizations.

We are still very interested in the impact the larger collaborative economy organizations have on society and politics as a whole. It's not like, "We don't want to work with Uber or Airbnb." It's more that we want to focus on startups or nonprofits that are still small and have the opportunity to grow and scale in a more decentralized way. That's why we are a big supporter of the open-source movement. We think there is an important opportunity to think and do things differently.

DP: One of the key things with many peer-to-peer organizations is this kind of blurring between movements and platforms with things like digital cooperatives. OuiShare and Shareable are examples of organizations that also function like platforms, which is why I like the term "people-powered platforms" for these next-generation, open network organizations that bring people together and create leverage through technology to affect real social change.

AL: I think "people-powered platform" applies well to OuiShare. First, it's clearly people powered, both in terms of governance and on the power people have in the organization. It also aims to be a platform or an incubator of people and products. People join OuiShare and become entrepreneurs or they start an initiative.

They can be local leaders, startup entrepreneurs, writers, or even creators of a local union for Uber drivers, which is a project we're working on in Paris. What matters is that they are driven by a set of values to make society work for everyone in a more inclusive and fair way.

What matters is that they are driven by a set of values to make society work for everyone in a more inclusive and fair way

DP: You talked about ending hierarchies, and I think the collaborative economy has a lot of potential for emerging markets, especially when billions of people will be coming online for the first time in the next ten to twenty years. What are your thoughts on the collaborative economy in a more global perspective?

AL: I think we need global platforms that people can use to scale their projects in a different way. One of the big visions behind OuiShare is to have Connectors all around the world that are knowledgeable about how innovation works in every city. Then you can scale a lot of meaningful initiatives in a different way. Policymakers and governments are really far away from all of this, and by spending time with them and sharing what we learn, how we work, and the way we operate, that can have a big impact.

Big corporations, startups, grassroots innovators, and local governments don't always talk to each other. Through OuiShare Fest and the online platforms we design, our goal is to create spaces where those people can have meaningful conversations. It's not an easy task because they don't have the same values or the same control codes, so it takes some experience to design spaces for enabling those conversations.

DP: I agree 100% that big corporations, startups, grassroots innovators, and local governments don't always talk to each other. A big goal of this book is to bring these various perspectives together in the hopes of starting a global conversation about collaboration, sharing, and prosperity for everyone. That is also the idea behind the title Empower: How to Co-Create the Future We All Deserve.

This also draws attention to tensions within the collaborative economy that we touched on earlier. When I attended OuiShare Fest in 2014, I noticed a lot of the conversations focused on ideas and values, whereas collaborative economy startups in the U.S. are heavily focused on metrics, revenue, and raising investment to scale very fast. This goes back to what you said about why OuiShare prefers the term "collaborative society" over "collaborative economy." How do you find a balance between these?

AL: Clearly, there's a big difference in the approach. Americans are much more pragmatic and interested in designing solutions. Theory matters to them to the extent that it can be translated into practical outputs. Europeans, and the French especially, are pretty much the opposite. Someone who gets a lot of authority in France is someone who can speak and write well. It's not someone who delivers practical solutions. It's about change and vision of society. I think the two approaches should feed each other.

We spend a lot of time talking, reading, and thinking about how to build the next generation of purpose-driven organizations. OuiShare Fest has become one of the main places where people can share ideas about theoretical impacts or emerging projects in that space. This enables us to have a unique vision that makes us more willing to support bold and radical ideas instead of something like the next Uber for X. That type of approach doesn't need us.

DP: One thing that comes to mind here is that Americans are often too focused on the short-term at the expense of the long-term. The first generation of unicorn collaborative economy startups coming out of the U.S. all took on so much venture funding that it is questionable whether they can sustain their market advantage. These startups may also have a devastating effect on the U.S. economy in terms of hollowing out good jobs, whereas Europeans are more conscious of maintaining a good quality of life and doing things that benefit the interests of everyone.

I'm curious what is the perception of U.S.-based collaborative economy startups around the world. Do you see tensions of them competing with local collaborative economy startups or disrupting industries, such as what Airbnb and Uber have done? Or does this go back to approach in terms of what you said earlier regarding traditional vs. radical and new?

AL: I don't think it's so much about U.S. vs. Europe or the world. It has more to do with how they are set up, how money is raised, and the values established by the founders. We try to share this with national governments.

National governments are still in the mentality of nations vs. nations, France vs. U.S. or whatever, which is old-fashioned. It's more about if you're a company in the collaborative economy space, then you should operate in a different way. There's an interesting framework from Jeremy Heimans of Purpose on old power vs. new power. We expect collaborative companies to be new power companies. That means new power values around sharing, collaboration, decentralized power, and some of the things we talked about earlier.

Uber and some of the big venture-backed companies in the collaborative economy space operate with old-power values. This creates risks because it creates expectations that the new power companies should also take on lots of venture funding and do things that undermine their new power values. Especially if you're a community-driven platform, you cannot operate the same way as the worst of capitalism used to operate.

We think the moves from Kickstarter, Etsy, and a few companies here in France are really interesting. They became B-Corps, which basically says to their community and stakeholders that they try to respect a set of rules and principles that make them different. To me, that's where the main difference lies.

It's not so much about U.S. vs. Europe. For example, BlaBlaCar, Drivy, and The Food Assembly are all global leaders in the collaborative economy space based in Paris. It's great to have European unicorns leading the way globally, but it's also great to have companies operating in a different way.

DP: This makes me wonder how the collaborative economy continues to evolve and be a people-powered movement when there are these massive investments of venture capital. Investors will inevitably expect massive returns. Larger companies are also acquiring local competitors or putting them out of business, suggesting a kind of consolidation phase similar to what happened with the rise of social media. Do you think the collaborative economy could evolve into some type of distributed network, similar to the way that OuiShare is organized? What do you see the future looking like?

AL: My main argument is that the collaborative economy is not like social media. I don't think it is going through this consolidation phase like social media did, where you had lots of social apps and startups, then slowly everything became consolidated by a few big companies like Facebook, Twitter, Google, and Apple. A lot of these unicorn companies are also probably unsustainable. I mean, they raised so much money and they will need to give a massive return to their investors. I don't see how that is possible over the long term.

The collaborative economy will continue to grow and evolve. The types of companies and movements that we support at OuiShare will be the next generation. They are the future. To build a collaborative society, we need collaboration and the people to be at the center, not investors and venture capitalists. That doesn't mean there is no place for investors because raising money can help them to scale, but rather that companies should be accountable first to their community.

I'm very interested in what Union Square Ventures is doing. I think blockchain is going to be a main disruptor and a great enabler for more decentralized models to happen. Uber, Airbnb, and BlaBlaCar are leading the way and are amazing disrupters. They're going to enable new models to operate in the future. I don't know when it's going to happen, but I'm sure that there will eventually be disruptors to this first generation of unicorn companies. Disruption is inevitable in every industry.

People tend to think that a company gains a monopoly and then the game is over. That's it. They think all of these unicorn companies disrupted traditional industries, and now they will last forever. But that's not how the world works. As Jeff Bezos said, one company's margin is another company's opportunity. These big unicorn companies take too big margins, and people are not going to use them forever.

People tend to think that a company gains a monopoly and then the game is over. That's it….But that's not how the world works

DP: *Another thing that comes to mind is that young people don't like to be on platforms with their parents. For example, part of the big exodus of people under 25 from Facebook to platforms like Snapchat was that their parents were joining Facebook.*

When I look at OuiShare's commitment to knowledge sharing, grassroots communities, local events, your ideals and values, you appear positioned to be a leader in the evolution of the collaborative economy in the next generation. As we look into the future, do you see any generational trends with the collaborative economy? What do you think young people are doing now that will become mainstream in the next 10–20 years?

AL: Yes, you said it. Operating with new values is going to be the mainstream. New generations want new values. When Generation Y or Z or whatever you call them comes to power, nothing will ever be the same. I'm pretty convinced about that. They have the values of the new power companies that we talked about earlier. There are few of these people at the top of big companies or with political power. When this happens, decisions that are going to be made will be very different than the ones today.

DP: *Today we are seeing a lot of nationalist, right-wing political movements gaining stronger support in Europe and around the world. There is talk about the EU, Brexit, Trump, the refugee crisis. As we think about the idea of a collaborative society, what role might OuiShare play at a national or international level? With your strong grassroots community and organization, could you imagine OuiShare becoming involved in politics or helping to shape economic policy in the future?*

AL: Of course. To what extent and with which design, I don't know. It's part of the ways we can make big changes happen. The challenge is that right now, if we had to do it, we would compromise ourselves because the people that are in power operate in a horrible way. We are waiting for big changes to happen, and we will support them when they do. But I think it's too soon. We need probably between 5–10 years for a generation to leave power. In that time frame, the values and things that we discussed will become more mainstream and we can start operating differently.

DP: I agree that the next generation will bring significant change, not only politically but also economically. Buying habits are changing. People want access vs. ownership, and prefer to work for or buy products from socially conscious companies. They are sharing and collaborating more, and all of that contributes to creating the type of collaborative society that you aspire towards. What does the future look like to you?

AL: I think the future is about decentralization. That means a decentralized economy and decentralized power, enabled by decentralized tools and decentralized or collaborative culture. What the collaborative economy is today is not the future because the way it is designed right now is not sustainable. It's just the same as the non-collaborative economy in the sense that it is focused on short-term profits, rapid growth, making money for investors, and these types of things.

> *What the collaborative economy is today is not the future because the way it is designed right now is not sustainable*

DP: This reminds me of Timothy Wu's book The Master Switch. *The telephone, radio, television, and cable were all invented by people with optimistic views of what the world should look like. Then as they evolved and money got involved, their inventions turned into these dominant monopolies. That gave rise to the Internet and blogging. There's ways in which when a monopoly forms, something forms in reaction to it. This cycle repeats itself over and over throughout history. Every time another monopoly gets dominant, we think the cycle will stop. Except it never does.*

AL: Yes, exactly! I think for the first time in history, we could have decentralized autonomous or decentralized collaborative organizations that cannot be centralized by design. That's the big difference with what happened before. The design principles were not established at the beginning. Things like collaborative decision-making, collaborative finance, and all kinds of peer-to-peer tools make it more efficient to scale an organization in a decentralized way than in a centralized way. When it becomes more efficient, then that means those decentralized powers get a way.

ANTONIN LÉONARD is a social innovator, collaborative economy specialist, and digital strategist. Co-founder of OuiShare and OuiShare Fest, he researches, consults, and speaks and on the power of collaboration and communities in a networked age. Antonin is a leading source of expertise for businesses and governments that want to embrace the collaborative economy.

IT'S A SHAREABLE LIFE

Chelsea Rustrum

Chelsea Rustrum is passionate about sharing and bringing people together. She advocates for action: Stop talking about the latest technology and do something meaningful. Join a co-op. Rent your spare room. Share your car. Go to a clothing swap. Meet your neighbors. Be part of your community. Create a society that empowers everyone to enjoy *A Shareable Life*.

- How the sharing economy meets basic human needs
- The world we want to live in isn't about technology
- The sharing economy evolution and business cycles
- It's important to be careful about what we are creating

The sharing economy began as a community focused, values-driven movement, which today is being eaten by on-demand everything. Our obsession over massive valuations and short-term gains jeopardizes the underlying mission to live in prosperity and abundance. Let's leverage technology to uplift humanity and build cooperative communities.

DP: *The title of your book is called* It's a Shareable Life. *Let's start with a simple opening question: What does it mean to live a shareable life?*

CR: Living a Shareable Life means to think about everything that you have and have access to in a way that can be shared: Your car being idle 90% of the time, a couch or an extra bedroom that you could open up, all of your stuff like clothes and books, how you eat, and how you get where you're going and the ways to do that with other people. It's also thinking about the things that you own or perhaps have access to. Do you need a car or can you use public transportation or one of many ride-sharing apps?

Sharing is about connecting your stuff to relationships in order to live a more simple and meaningful life. There are many extrinsic benefits. People tend to save or make money when they share. Sharing also provides a greater context for human interaction, prevents isolation, and gives people more stability when they can sense their place in a community. When you visualize your life in terms of what you can share, you naturally realign with what it means to consume vs. have access, and what is a liability vs. an asset. Ownership takes on a whole new meaning.

The book is a practical guide to sharing with plenty of stories and antidotes, but it's important to find a place to start and experiment with what works for you based on your interests and comfort level. If you love to cook and maybe someday want to own a restaurant, have strangers over for dinner. If you enjoy being hospitable and nurturing others, have a guest and see how that feels. If you own a car, try offering access when it is idle. If you are into fashion, bring a few items to a clothing swap. Start with what excites you and see what happens.

DP: I'm a former scholar of religion and culture, and I used to do a lot of research on the birth of the modern world. One of the things that I love about It's a Shareable Life *is that you open with the discussion of how people used to share a lot more than they do now. Even though we've seen an increase in wealth and prosperity in the last fifty years, with average house sizes more than doubling and incomes growing dramatically, over 25 percent of the population suffers from diagnosable mental illnesses such as anxiety and depression.*

You suggest that this may have something to do with going from a shared economy to an individuated economy. This raises a lot of interesting questions about values. What do you think happened and how is the sharing economy in a way a kind of return to how things used to be?

CR: We seem to be in transition, where adoption of sharing practices and platforms is reaching the masses, but there's a fine line between sharing and the efficiency of straight up convenience. There's a real beauty to people opening their homes, allowing their neighbors to drive their cars, treating strangers like friends along the way. I think we're feeling disconnected with all of our technology and individual apartments and single-family homes. In cities like San Francisco, people have their own stuff and can easily get their laundry done with an app, have food delivered within 30 minutes, and receive massages on-site, but all of that convenience can come at the expense of not knowing who lives next door and not contributing to your local community.

All of that convenience can come at the expense of not knowing who lives next door and not contributing to your local community

I often like to think about what we actually need as human beings. I contend we need love, belonging, food, shelter, and creative expression. We've created a world with a lot of stuff that we don't need and we can be seduced into believing that those are needs instead of just desires. At a certain point, we can't continue buying, buying, buying. We only need so much stuff. As a society, continuing to grow bigger and bigger doesn't make sense and won't work. So the question becomes—What do we need and what's the best way to arrange a society around that?

DP: *This is what Douglas Rushkoff refers to as the growth trap.*

CR: Yes, exactly. It's about greater efficiency, so a lot of these platforms focus on removing friction to make sharing and collaboration easier. But I think there's a step beyond that which is where we started in the sharing economy. How do we connect at a greater level? How do we get our needs met both collectively and individually and maybe even do that at the same time?

This movement of sharing has a long horizon that will need to be written into our social structures at a deeper level than apps can provide.

The focus is less on "me" and more on "us." I firmly believe that this is just the end of the beginning, moving toward shared ownership, shared value, and a new economy or new paradigm that looks very different. If we serve economic instead of human needs, eventually that has to crash.

DP: This sense of disconnection and emphasis upon the individual over community is a modern phenomenon. It reminds me of my research at Princeton on the origins of self-help. In the 1800s, etiquette books became very popular in part because people basically had to learn how to walk past strangers. There were mass migrations from rural areas to cities and people weren't used to ignoring others or making judgments based on social or economic status.

Earlier you touched on the significance of access to knowledge, resources, tools, and services. In the book, access is the foundation of progress in the sense that it is an enabler to empower people with opportunities for more meaningful lives. Yet we continue to operate under an ownership model where things are owned by a single entity. Tell us about access vs. ownership and how access models relate to the sharing economy?

CR: Everything is on loan in the sense that you can't take things with you when you die. We have models like "this is mine" and "that is yours," so don't touch it. Our economy is wrapped up in these images where ownership is associated with status. Ultimately, what we want is access to stuff when we need it. Not having to own things and sharing resources frees us to live with more meaning and purpose. That was the premise behind the idea of my book *It's a Shareable Life.*

The challenge is that the sharing economy started as a movement and is becoming an emblem for a business model. It takes only a small step back to realize that many leading companies in the space are absorbing the value of assets and time that people exchange on these platforms. Wealthy investors and venture capitalists that back most of these startups expect generous returns on their investment, which will create greater economic divide.

Right now, on-demand services are eating the sharing economy. The next iteration of this will be the rise of the autonomous world with things like drones, self-driving cars, and the IoT (Internet of Things). This will happen at the expense of peer-to-peer connections becoming less important in part because the autonomous world will be fueled by massive investments of venture capital and big corporations.

We need to look at access vs. ownership in the context of who owns the platforms. The technology powering the sharing economy is not expensive to build, and yet these companies have skyrocketing valuations based on the value of other peoples' assets and time. There is a way to more equally distribute wealth and transfer value. For example, we can easily build technology to account for, assign, and distribute value as it's created.

We need to look at access vs. ownership in the context of who owns the platforms...to more equally distribute wealth and transfer value

Value distribution is coming, but it needs to start with the decisions made by founders. For example, things like sharing ownership or creating terms that favor users instead of investors. This will ultimately create greater access for everyone.

DP: One thing that really excites me is that we're the first generation in history to go from an analog to a digital world. There is this fundamental, irreversible shift happening right now for human civilization and the sharing economy is laying the foundation for a new way of being in the world. Given the challenges that we just discussed around access vs. ownership, what is your vision for a shareable life as we go into the future?

CR: I think technology is the less interesting part. Right now we're at the point where technology is ubiquitous. It's replicable and a commodity. For example, it's still difficult to build these things, but you can pay someone twenty thousand dollars to build a look-alike Kickstarter. The barrier to entry is low on the technology side.

What I see is people working on new business models and trying to harness them or use the technology that we already have.
I call them digital cooperatives because they distribute the value
through sharing online. Sharing goes through the very structure
that teaches us where value comes from. Sharing becomes the
incentive that drives us. And instead of building marketing in as
an add-on, the community is the company.

For example, people work so they can earn enough money to pay
their mortgage. But what if we reframe housing and build connected
communities where people share costs? What if you don't have to use
an app every time you want to share a car or rent a room on Airbnb,
but instead it's just built into how we live? This is a kind of rewiring
of our inner technology. It's a cooperative model for everyday life.

*DP: On the one hand, "on-demand everything" is eating the shar-
ing economy in the sense that the sharing economy is being equated
with a new type of on-demand capitalism. This is happening at the
expense of overshadowing peer-to-peer interactions and connections.*

*Then there is this idea that the technology is the least interesting
part. The sharing economy is about sharing. It's a people-powered,
community-driven movement with all of this incredible potential
to revolutionize every aspect of our daily lives. People are sharing
more and more, leveraging existing technology to create digital
cooperatives and new business models.*

*It's as if there is all of this momentum to force technology and cap-
italism to fit into this on-demand everything framework, yet at the
same time there is a growing awareness and adoption of a sharing
economy movement that is decidedly egalitarian and democratic,
driven more by community than by capitalism. In fact, this movement may lead to value distribution and decentralized ownership,
disrupting venture capital.*

*Can you tell us more about the sharing part of the equation? How
do you see the sharing movement continuing to unfold?*

CR: Business goes through a predictable cycle: idealism, adoption,
and integration. The same patterns repeat over and over. The sharing

economy is no different. We simply haven't integrated sharing yet. This is what I meant earlier when I referenced this idea that we are sort of midway through the adoption of sharing practices.

While the sharing economy seems to have lost much of its original meaning, we now have a social movement that is full force, breaking down the social constructs of ownership and exchange. People are thinking more about the value that they already have, considering access over ownership, integrating sharing into their daily lives, and slowly this is creating a new consciousness around abundance.

> *While the sharing economy seems to have lost much of its original meaning, we now have a social movement that is full force*

This type of thinking is also coming into venture capital. For example, Fred Wilson of Union Square Ventures and author of the popular blog A VC, recently wrote, "As more and more businesses leverage the power of networks to create economic value, there is a question of whether the network participants should share in the value they help create." In this way, we are starting to map the possibilities of an economy that shares the responsibility and value with its participants.

DP: *The thing about movements is that they change behavior and shape expectations that the general public has from larger institutions. As the sharing movement continues to build and grow into a mainstream phenomenon, I would expect venture capitalists and larger corporations to explore how to integrate sharing into the fabric of business in order to meet consumer demand and remain competitive.*

What are some examples of integrations that could become common as the sharing movement continues to grow?

CR: We need to start thinking about the type of world that we want to live in, and that will impact how we integrate sharing. For example, if you think about the autonomous world, we need to consider how the ownership of machines will impact everyone and build models that will ultimately share wealth because we don't want to live in a world controlled by machines and a few wealthy people.

There are many different ways to structure organizations to distribute value. We already discussed cooperative models, for which there can be multi-stakeholder approaches allowing for outside investment. Then there are ESOPs (Employee Stock Ownership Plans) that can distribute value among value creator employees. Unlike cooperatives, the benefit of that approach is that the business can be sold.

Cooperatives can also be utilized in digital environments to create what are referred to as platform cooperatives. Crowds can act like a marketing engine if they are given the proper incentives, driving growth of platforms where users share in value creation. A few examples include Stocksy (a stock photo marketplace), Fairmondo (a more conscientious eBay based in Germany), Lozooz (a blockchain version of Uber), and Loomio (a group decision-making tool used by organizations like OuiShare.)

There is also a lot of potential for the blockchain, a technology that records exchanges to create more secure transactions. Sharing is about peer-to-peer exchanges, and a truly peer-to-peer system built on the blockchain would allow for transactions to happen without the need of an intermediary platform. It can also be utilized to create self-reinforcing contracts and make it possible to facilitate transactions between strangers without the need for lawyers, accountants, and bankers.

Crowdfunding led by companies like Kickstarter and Indiegogo has also become an explosive industry. Recently, legislation was passed that allows crowdfunding of equity in companies. If this trend continues, crowdfunding could become a worthy alternative to venture capital, especially among entrepreneurs that want to build new types of peer-to-peer companies that distribute value.

For example, there are emerging platforms like Republic and Wefunder that give anyone the ability to invest money in start-ups and existing companies in exchange for ownership equity.

Imagine a reality where companies that aren't publicly traded or even formed yet can make their users, members, providers, and participants part owners.

But we need to look at all of this together. We haven't fully realized the power that individual transactions have. Our individual purchasing power as a collective can directly impact all industries, leading to an improved quality of life for everyone.

Imagine a reality where companies that aren't publicly traded or even formed yet can make their users, members, providers, and participants part owners

DP: *One of the challenges in talking about the impact of the sharing economy movement that you describe is that it is so ubiquitous and wide reaching. There is the groundswell from the bottom up of peer-to-peer sharing, and then this kind of top-down impact of things like platform cooperatives that could disrupt traditional industries and even disrupt the first wave of sharing economy start-ups funded by venture capital.*

When I think about integration and the different layers of sharing that you describe, it's as if the sum is greater than the individual parts. I've heard you describe the sharing economy in a similar way. Can you elaborate on this and help us understand where you think the sharing economy movement is heading in the future?

CR: It's a mentality shift, a cultural shift, a movement from "me and mine" to "us and me" and defining a space where all of that comes together. The power of the individual in combination with the power of the crowd is the superpower. There is a kind of realization of individuality in the context of being part of the collective.

The big, venture-backed leaders of the sharing economy like Airbnb and Uber are necessary just like AOL and Compuserve were necessary in providing original entry points to the Internet. I don't think the future of the sharing economy will look much like what it does today, and it is important for us to decide what it will look like together—as a collective.

We need to be careful about what we are creating. It's important for people to stand up for the integrity of the movement. There is this sense of a renewed faith in humanity that millions of people share feeling connected to each other, sharing what they have access to, and doing things that are mutually beneficial.

We need to be careful about what we are creating

Businesses will need to consider the value of sharing beyond the typical rationale of needing "ambassadors" for marketing purposes. They will have to find creative ways to compete with new and copycat sharing alternatives that will empower people to do more with less. Once sharing is built into the financial models, legal structure, and the very backbone of our economy, collaboration and cooperation will be a foregone conclusion.

DP: We talked about the idea of a shareable life, building a sharing economy movement, and the focus on peer-to-peer cooperation, collective actions, and value distribution being more important than technology. All of this points to a sense of purpose where sharing helps to actualize or manifest the potential hidden within us all.

Through sharing and participating in something greater than ourselves, we are able to realize our true potential, which in turn allows us to contribute more for the sake of the greater good. I love the utopian optimism of this vision, particularly how it stands against the dystopian view of a world dominated by machines and controlled by a handful of wealthy elites.

As a final question, how do we continue to live with purpose and make sure that the sharing economy retains its integrity and values? To draw upon the title of the book, how do we empower each other and co-create the future we all want and deserve?

CR: I live in San Francisco, and being there puts a lot of pressure on people. There is always this sense that you are never "enough"—

rich enough, successful enough, unique enough, social enough, etc. This leads to lots of criticism, name dropping, fear, anxiety, uncertainty, and so on.

I think we need to focus on being happy and alive in this moment. Let go of the never-ending comparisons and chasing money or perceived success. Be comfortable with our flaws, surrender to our aging bodies, and let go of this idea that you have to earn your right to exist. Maybe it's enough to be alive. If we start from that place, then we can approach each moment with compassion, mindfulness, and that will shape our relationships.

The social transformation of "enoughness" is key to evolving the economy. All of the innovation and creativity is great, but to create the new structures in the material world that we need in the future, we must start as locally as we can—with ourselves.

....

CHELSEA RUSTRUM is a sharing economy author, facilitator, and consultant with deep practical knowledge and hands-on understanding of how the collaborative economy has come to fruition, grown, and continues to change business and disrupt industries. She's the author of *It's a Shareable Life*, the founder of a social and educational series in the Silicon Valley, dubbed *The Sharers,* has advised dozens of marketplaces, and speaks to corporate audiences such as PwC and J&J, helping organizations understand the changing nature of business. Chelsea has participated in various conferences and summits speaking on the growth and development of the collaborative economy, including TEDx, Expo Italy, Grow Co., and has also contributed to articles in *Inc. Magazine, Wall Street Journal, Wired, Forbes, and The Economist.*

....

PART III

COMPANIES AND THE CROWD

*How can you co-create
solutions with the crowd?*

*What strategies are needed
to adapt and innovate?*

*Why does purpose drive
success and growth?*

SMARTCUTS

Shane Snow

Shane Snow is a passionate storyteller who revolutionized how brands tell stories. He went from a struggling freelance journalist with at one point only $0.48 in his bank account, to co-founding Contently, a leading content marketing platform powered by a community of 55,000 journalists with clients including *American Express, General Electric, GM, PepsiCo, Coca-Cola, Google,* and *Walmart.*

- Building a business based on the needs of the crowd
- Why content can be more direct than advertising
- Asking smart questions leads to breakthroughs
- Great innovators stand on the shoulders of giants

Contently pioneered a new crowd-based business model that Arun Sundararajan calls a hybrid services-marketplace, where a company provides services to enterprise clients that are crowdsourced from a marketplace of providers. Shane remains a journalist at heart, writing his bestselling book *Smartcuts* while he built Contently. Packed with insights and great stories, this interview is a pleasure to read.

DP: There is this idea with content to start everything with a story. You have a great story of having 48 cents in your bank account at one point before raising money for Contently. Why don't we start there?

SS: Sure (laughs). Three of us had this idea to start a company. One had a good amount of savings from his last startup—that's Joe, our CEO—and the other had six months before his student loan payments started. He also owned his own place. I had credit cards and not really any savings. We basically had a six-month window to prove that there was an opportunity, business potential, and that we were the people to build that business. Then we needed to start making money or raise money.

The 48 cents in the bank account was right at that point when we figured out a lot of things about the freelance economy, where the supply and demand were, and what was going to work for Contently to turn it into a real business. We started making a little bit of money, but not much, and we had been courting investors. It came down to the wire: "We need to raise money soon because we're going to be broke and the dream might slip away, but we're past the point where it's totally ludicrous to invest."

I took a screenshot when I checked my bank account and it had 48 cents. It was kind of a panic moment, but then I had the foresight to take a screenshot so that I could remember later, no matter what happened. We ended up raising money soon after that, so I was able to start paying rent. But yes, times were tight.

DP: *We met after Contently got into Techstars in 2011, and that was before most marketers or investors were familiar with the concept of content marketing. What was the initial vision, and what was the journey like for you and the founding team?*

SS: The initial vision was to solve problems for freelance journalists and the companies that hire them. Everyone that I knew in journalism was graduating with no job, losing their job, or looking for a job. Basically, we were all hustling as freelancers.

There's a whole list of problems, or challenges, that go along with being a freelancer—in any field, but in journalism, it was pretty severe: finding work, getting clients to pay you on time and consistently, marketing yourself, doing your taxes as essentially a small business owner, getting credit for your work, etc. How do you help these people whose skills are in demand, but who are on their own? How do you help them deal with the business side of being a freelancer and a creative?

How do you help these people whose skills are in demand, but who are on their own?

The other side of the problem was that companies still demanded those skills at a level that you weren't finding on typical online talent labor marketplaces, like Elance, where it's more about outsourcing commodity labor, and less about highly paid creative labor. Initially, Contently was about matching those groups up—people who wanted to pay more for creative talent as freelancers, and freelancers that needed help running their business, getting work, and getting paid on time. That was the initial business.

We were platform agnostic regarding who the buyer was, but the theory was that our main customers would be media companies—newspapers, magazines, and the digital arms of NBC and CBS. It wasn't long before we figured out that commercial brands had the biggest, most acute need for these people to do content marketing. They were blogging and trying to put content out on social media, but they couldn't make it work with Elance or their staff, and agencies were too expensive or were more interested in doing television commercials and big projects than web content.

We still do some work with media companies but for the most part we went all in on content marketing. This basically turned Contently into a software company, where we build analytics, project management and workflow tools, integrations into Marketo, and all sorts of things like that to streamline content marketing based on the needs of big brands. But the initial core vision remains the same, which is you need high-quality, talented, creative people to help you make good content and tell good stories.

DP: I want to dig a little bit deeper into that. Contently has grown to over 100 employees with 55,000 storytellers and journalists, and 8-figure revenue with an impressive list of top-tier brands like American Express, General Electric, GM, PepsiCo, Coca-Cola, Google, Walmart, and so on. If you started off solving problems for the storytellers and journalists, what problems did you need to solve to scale and meet the demands of the brands?

SS: Content marketers or brand publishers basically go through a step-by-step process. First, you need a content strategy. Second, you need ideas for what you're going to write, shoot, and publish. That is the ideation part. Third is actually creating and managing content.

Creation is not only who's going to create it but who's going to approve it, how do you (especially if you're a big brand) track who touched what so that you can go back and audit things if something goes wrong (lawyers have all that), how do you make sure that the lawyers, PR people, and whoever else needs to be involved in approving content are always in the loop, etc.

Then there is the actual publishing of content. You have to put it places where people are going to find it, you may promote content (so, advertising it or basically using media buys to get people to your content using things like Outbrain or Facebook-sponsored ads), then reporting and measurements, and finally, learning from all of that so you know what to do next time (what we call "optimizing").

Successful content marketers were largely doing this manually either internally or with agencies, and many companies coming to us couldn't do anything with just the talent. They needed the workflow and process to be able to manage the internal bureaucracy of approvals, publishing, reporting, and measurement. We basically baked the entire creation process into our software.

Our freelance network, in turn, got to this point where we no longer had a problem with supply. We had all these people who want work and all of this data about them, so we can match a brand with very specific, vetted, and talented people. In order to keep them on our platform, we had to provide value by helping them to find work, get paid, do taxes, market themselves, and manage their business.

We realized that we could actually do our own content marketing on those topics in order to keep our community happy, engaged, and associated with Contently. We have a blog, a huge mailing list, daily newsletter for freelancers, and free tools. Taking care of our community allows us to build stuff for the brand, since the throttle on how much work we can get our freelancers is based on how many brands we have as clients doing content marketing success-fully enough to want to do more of it.

DP: Contently is a great example of how brands can collaborate with the crowd. In this case, it's a crowd of high-quality con-tent creators. Contently is also a flagship example of what Arun Sundararajan referred to in our interview about crowd capi-talism as a hybrid-firm marketplace, one of the new emerging business models for the future of work that combines freelance marketplaces with enterprise-level services typically found in large firms.

Can you reflect a bit on what you think is innovative about your approach to managing the relationship between companies and the crowd? What lessons might other companies learn from what you've done?

If you are not providing for the crowd, then they're not going to do any work for you and they won't help you out

SS: You need to provide value on both sides. You can't just think about how to make money from your clients. If you are not pro-viding for the crowd, then they're not going to do any work for you and they won't help you out. We learned that by giving first— by putting stuff out there to help freelancers, by building them free tools to market themselves and build their portfolio without asking anything of them—we engendered a lot of good will. That helped build our reputation, get users, and create marketing that drives the business—but it also makes them more likely to look at an offer from our brand clients who want to work with them. They know that they're not just mercenaries for hire.

DP: The idea of "giving first" makes a lot of sense, especially since you are a Techstars company. I spoke with Brad Feld about this. Can you dig deeper into a few examples of how to provide for the crowd?

SS: The payoff needs to be immediate. The biggest surprise to us was that getting people paid early and on time is more important than how much they get paid. We try to keep rates high, but people are willing to bend over backwards if you're going to pay them on Saturday instead of three months from now. This is a big deal in the freelance economy and when you're working with a crowd. If the payoff is immediate, then you can get a lot of people to collaborate and jump through hoops that you might normally need to pay more money for.

You should also think about clever uses of the crowd. For example, look at the CAPTCHA project to digitize old books. They didn't ask people to digitize text, because that is boring and uninteresting. Instead, what they did is ask you to fill out a CAPTCHA to authenticate you are a real person. Instead of paying people 10 cents or something per CAPTCHA, they motivated the crowd to complete tasks, not realizing that they were actually transcribing books.

You can't make people take a gamble on something that they don't know or can't see. That again goes back to the idea that the payoff needs to be immediate. Your relationship is pretty tenuous because they are the crowd. There needs to be this kind of instant gratification component, which may be isolated or removed from the bigger picture like with CAPTCHA, where they transcribed text and were instantly rewarded with an eBook download or access to a website.

On the flip side, the more that you can make people feel like they have a relationship with you, the more they'll go to bat for you and the better work you'll get. For example, we started with this sort of crowdsourcing paradigm: Let's get as many people as we can, let the good ones rise to the top, and then filter out the rest. It was a "may the best man or woman win" sort of thing. We'll get the best quality that way.

What we found is that things got better when we treated people personally or made work feel personal. For example, things like personalized emails and once in a while a talent manager calls you on the phone. Have in-person events where people can see that we're human and meet the people that are hooking them up for work. Then word spreads. Even if only 1% of the people we're working with ever talked to us, we give them such a good experience that they're going to tell other people.

I think many companies are great doing that for paying customers, but you need to apply the same idea to the crowd. Yelp is a great example of this. They invite their top reviewers to exclusive parties where they meet them and hook them up with stuff. Make the crowd feel special by doing things that build relationships, because otherwise your brand is faceless and they won't care.

DP: Storytelling has always been important for brands, but clearly, something changed in the past few years with the rise of content marketing championed by crowd companies like Contently. In parallel, we've seen the radical disruption of journalism that you talked about earlier.

Thousands of media brands went bankrupt while others such as BuzzFeed had a meteoric rise to become billion-dollar companies, monetizing exclusively through branded content with no ads. If you were a journalist reporting from the frontlines, what would the story be around content marketing and the media landscape? Can you help us navigate the shifts that happened in the last few years?

SS: Media, stories, and content were always subsidized, usually by brands. In the early days, we paid for the newspaper and magazines. It might be only a couple bucks, but that's not the whole cost. They made their profits from advertisements. As a brand, you were subsidizing the stories because the newspapers, magazines, and everyone else had the audience that you wanted to reach.

Today, brands can make that content instead of simply advertising by it. You used to need a staff of journalists, a printing press, and a fleet of trucks, or you paid that company for an ad. Today, companies like ours can connect you to freelance journalists all over the Internet, the printing press is WordPress, and the trucks are social media. Across any category—entertainment, education, general information—companies see it better to create stories that people want rather than buy ads next to them.

This basically gives them more control over the branding and advertising experience, especially if you can get people to come to your own website to read the story, or if you can get them to open your email. That is what shifted. It leaves a bit of a question mark for the hard news, investigative journalism, that orphanage fire story. How do we pay for that if brands are doing content elsewhere?

> *People's attention is getting so split that it is a smart strategy for brands to build their own audience directly*

The other big thing that happened is there used to be a limit to the places where you could reach people. This is why media companies could charge so much money for commercials and ad space. The Internet provided unlimited space, and especially with companies like Google and Facebook, you can reach people more directly in a more targeted way than before. This basically means that ads cost less on media sites and they get less attention.

As a result, people's attention is getting so split that it is a smart strategy for brands to build their own audience directly through things like email and social media rather than rely on side sources. Content marketing becomes cheaper and more effective than competing with traditional advertising spots. To summarize, there is more stuff out there, and brands can now make the stories and give people the media and entertainment that they want. The basic idea is get people to know who you are and pay attention to you.

DP: I'd like to switch directions and turn to your best-selling book, Smartcuts. *To start, what is a "smartcut"?*

SS: Smartcuts are about rethinking the way things are done, or what psychologists call "lateral thinking." It's approaching problems from unconventional angles or taking things on by rejecting the rules and conventions that are already there. The term "short-cuts" is amoral. You could take a shortcut through the woods and it's fine, doesn't hurt anyone, or you could take a shortcut that brings the financial industry down, and that's not good.

Any time in history that you've seen breakthroughs happen—whether it's in business or the arts or science—it's because some-one has fundamentally rethought the way things are done. A clas-sical example is gravity. Everyone took it for granted. Einstein was bold enough to rethink Newtonian gravity and that's how we got relativity and all sorts of breakthroughs. Being willing to question things that we assume leads to doing things in a new way. That's what *Smartcuts* is about.

DP: Rethinking how things could or should be done leads to breakthroughs by forcing us to break conventions. A key com-ponent of this process in the book is to ask challenging ques-tions that are impossible to answer by simply working harder. For example, "How might you make a 10x improvement?" or "How might you cut costs by 500%?" What types of questions should people ask?

SS: The kinds of questions that I like are ones where the question itself literally puts you in someone else's shoes. How would x per-son who's very different than me look at this problem? For exam-ple, the ones I always use are "How would a ballet dancer look at this problem? How would a firefighter . . . ? How would a racecar driver . . . ? How would a child . . . ? How would my grandmother look at this problem?"

When you seriously consider things from other perspectives (you can even go as far as asking those people what they think, which can get into the idea of customer development), that's a good way to help you realize what are the assumptions you take for granted. There are inherent rules in any industry or project, so then the questions becomes about the inherent rules themselves. "What are the assumptions inherent to this problem? If we changed them, what would happen? Does that lead us to something new?" I like those kinds of questions.

DP: When I think about the importance of questions, the next logical step for me is the idea of testing. There are all types of great digital tools that we can utilize to test and optimize. These allow us to eliminate guesswork and reach the right people with the right message at the right time.

There's a great example in Smartcuts *from Upworthy, where they sent a video to different subscribers with different headlines. After examining feedback and click-through rates, they found a 20% better click-through rate on one headline. Then they repeat the same process 75 times, pairing headlines and images, to eventually get an astonishing 69% click-through rate and 186% increase in video views.*

That leads to the question: Why is rapid feedback important? How can companies leverage their relationship with the crowd to test and iterate faster?

SS: That example connects to lateral thinking because in media it's assumed that once you publish a story, if it takes off, it takes off, and if it doesn't, it doesn't. It's like, "Is it a hit or not?"

What they questioned is: Is that actually true? And they said, maybe it's the packaging that makes a difference, and maybe you can repackage something. That was the lateral thinking, the smart-cut. The rapid feedback idea is similar in the sense that the faster you can go through the iteration process on an idea, the faster you'll get to whether it will work or not.

An analogy I like to use is to imagine that you are trying to land on an aircraft carrier. You have one shot, so you can't blow it. A lot of the way companies think about launching products and putting our ideas out there is to wait until it's fully baked and ready. Then we try and land on the aircraft carrier. Things that are worthwhile and important are hard, and so it's easy to miss the aircraft carrier. One tiny thing can go wrong, or everything can go wrong.

This is why they built flight simulators. You can practice landing on the aircraft carrier thousands of times without ever risking dying. In the flight simulator, you work your way up to harder and harder scenarios rather than starting by landing on an aircraft carrier in a real plane. You start off doing small tasks in an environment where if you die, it's just a game, so it's okay. That's the idea of rapid feedback. You test in a low risk situation over and over, so by the time you're in a high-risk situation, you've already ensured that you're going to be successful.

DP: When I think of how smartcuts relate to Contently, you made it easier for brands to tap into the crowd by radically rethinking the entire process. Only high-quality journalists are allowed on your platform, you make payments easy for the brand by not having to hire everyone individually, and then you pay journalists upfront to mitigate the risk. Each step bakes smartcuts into your platform.

How do you transition from smartcut thinking to building a startup or creating breakthrough innovations? Can you tell us a bit about the importance of platform thinking and leverage?

SS: Platforms and leverage are about lots of people doing lots of great work. As Newton said, everything that he owed was by standing on the shoulders of giants. You could stand on the shoulders of giants or you can be too proud or ignorant and try to stand on your own. But someone else who stands on the giants' shoulders will be taller than you. That's the idea. Look for places where people have already done work that you can build on top of, rather than reinventing the wheel.

The analogy that comes from my childhood is it's really tough to pry up a nail with a hammer. However, it's easier if you put a piece of pipe on the end of the hammer and make a lever. It's a mathematical law by Archimedes, the Law of the Lever. You can push down with less effort and get more or equal results. This is what your business should do—make something easier, cheaper, faster, or whatever for the same amount of effort that you had before. That's what you're trying to do as an entrepreneur.

This goes back to asking smart questions. In every aspect of your business, you should ask, "Where can we get more leverage? Where can we get more bang for our buck?" Give yourselves challenges that are so big that you can't solve them by working harder.

Force yourself to find leverage or be creative. At Contently, we periodically have meetings where the topic is "How do we make this thing ten times better?" We call each other out on ideas that are smaller than 10X, and that forces us to come up with bold and weird new things.

Starting from there is a much better place for innovation. There are a lot of different approaches you can take. Think about who has already done work, made the science, or built something that you can leverage instead of building from scratch. How can you get more results for a lot less effort, as much bang for your buck so that you can double down on something else? That's the idea of leverage.

Think about who has already done work, made the science, or built something that you can leverage instead of building from scratch

DP: In Smartcuts, *there is a cool YouTube case study of a one-hit wonder vs. a runaway success story. One was the double rainbow video by Bear Vasquez, which received millions of views after Jimmy Kimmel called it the funniest video in the world, and the other one was by Michelle Phan, a 23-year-old Vietnamese-American makeup artist who showed how to recreate makeup in Lady Gaga's video, "Bad Romance."*

The former became a one-hit wonder, whereas the latter went on to become the cosmetic queen of the Internet with 800 million views, 5 million subscribers, and her own makeup line by L'Oréal. As a final question about companies and the crowd: What lessons can be learned in terms of how big brands can tap into the power of the crowd, and how the crowd can collaborate with big brands?

SS: The first is the myth of the overnight success. A lot of us get our hearts broken when we build up to this grand moment and the grand thing doesn't go viral. The one-hit wonder is often random and hard to predict. Michelle spent a couple years making lots of really good content that served as potential energy for when she was discovered. This is what you see with a lot of people that seem to pop up out of nowhere, including inventors and business successes. There's almost always a lot of behind-the-scenes work building up potential energy before they become a success.

She was able to show people more when her date came and she was discovered by the masses. Once you're done with the video that you saw, it said, "Here's 50 more videos by Michelle." She was going after subscribers and fans, not just views. In contrast, Bear Vasquez was a happy accident. He kept pursuing one-hit wonder type of videos instead of trying to get true fans and subscribers.

Michelle also figured out how the YouTube homepage worked. She realized that she could get exponentially more leverage if she could get on the YouTube homepage, and the easiest way to get the most time on the YouTube homepage was to get a certain number of views to a video on a Friday afternoon. If that happened, then you'd stay on the homepage for the whole weekend because they didn't do anything over the weekend. Over the course of 50 videos, she built up a small audience of really big fans. She would post on Friday afternoon and beg everyone that she knew and all of her fans to check it out at the same time so that she could get that little boost and accidentally stay on the homepage over the weekend.

The strategy is to basically study how the system works and figure out the easiest injection point. Eventually after doing this for a while, someone discovered her on one of those weekends and wrote about it for *BuzzFeed*. This was the classic "someone with a big megaphone discovers something, it goes viral, everyone talks about it, and then it goes away"—except in her case, it doesn't go away. She had all of this potential energy there to capture people, and then she kept on building more stuff and parlaying that momentum into more things, including makeup lines and businesses.

For a brand, there are a few lessons. The main one is that if you want to be smart with content marketing and social media, then you need to spend a lot of time studying the environment. You can't do the same thing that everyone else does, and you can't try to recreate hits—yours or other peoples'. You need to aim for subscribers, and that comes from deep, consistent, high-quality content aiming for the right relationship rather than a mass widespread audience.

There's this great article by Kevin Kelly from *Wired* about having a thousand true fans. If you have a thousand true fans, then they can get any ball rolling, vs. a million people who just glance at you.

You can't do the same thing that everyone else does, and you can't try to recreate hits— yours or other peoples'

I think what a lot of brands do is they give up too early. They try and aim for a hit they see ("Oh, Dollar Shave Club did a funny video. Let's do one exactly like that."), not realizing that some of the success of that video is random chance and it's a better strategy to do consistent, high-quality content and build up to get those thousand true fans. Then, when their time comes and people do see their random viral hit or whatever, they'll have sustainability that they can then parlay into more.

SHANE SNOW is an award-winning journalist, celebrated entrepreneur, and the bestselling author of *Smartcuts: The Surprising Power of Lateral Thinking*. He is co-founder of the content technology company Contently, which helps creative people and companies tell great stories together, and serves on the board of the Contently Foundation for Investigative Journalism.

Snow's writing has appeared in *Fast Company, Wired, The New Yorker*, and dozens more top publications. He's been called a "Wunderkind" by *The New York Times*, a "Digital Maverick" by *Details*, and his work "Insanely addicting" by *GQ*—though he wishes they had been talking about his abs.

PURPOSE

Michael Bronner

Michael Bronner is the founder and chairman emeritus of Digitas, a global digital marketing agency employing six thousand people worldwide. Under his leadership, Digitas redefined modern marketing. Michael's next company Upromise helped 10 million families save over $70 billion for college, becoming one of the first major success stories of the collaborative economy.

- Building the most successful rewards program ever
- The power of leading from a place of consciousness
- Why the essence of good strategy is sacrifice
- Investors want to back founders that change the world

Michael's success stems from his commitment to serving the needs of others. He embodies the type of integrity and values that we need in large companies. Michael is also one of my favorite people in the world. I am grateful for the opportunity to work with him on his latest company UnReal, I am proud to consider him a mentor and friend, and I am honored to share this incredible interview with you.

DP: Your personal story sounds a bit like a fairy tale. In your junior year of college, you found and turned in a wallet of a banker who gave you a line of credit that allowed you to found your first company creating hundreds of thousands of coupon books for Harvard and other Boston schools. Eventually, you went on to corporate clients, including American Express and AT&T. Coupon books existed at that time. Can you tell us about your backstory, and what was revolutionary about what you were doing?

MB: I have a strong memory walking across campus one day and they were selling a big, fat coupon book for $10. I remember looking through it saying, "Hardly any of these I would ever use," and kept going. Then I thought, "Why should you have to pay for a book of ads?"

It's one of those things where an idea comes out of something that you see. I went to the Dean of Housing at BU and said, "I'm a student. I know what students want. If I create a coupon book with all local merchants ('Buy 1 pizza, get 1 free,' '25% off at this store,' access into clubs, things like that), would you put it in the mailboxes?" He said, "Yes."

I went to merchants and convinced them to pay for an ad, and I required them to do something very aggressive to drive business. In one day, 20,000 of these books were given to students in their mailboxes, and the merchants couldn't handle the volume. It was insane. That was the beginning. I was a biochemistry major. I didn't know much about business. It just became very clear, this idea of targeted, direct marketing.

In the next few weeks, I locked up 100,000 mailboxes at most of the universities across Boston and started selling the advertising into these books, and then I took the same concept into downtown offices. I figured they have all of these people going up and down elevators all day. There were hundreds of thousands of people in a three-square-mile radius. I went to these big companies like Hancock, Gillette, and Prudential and said, "If I create a coupon book with great offers from the local merchants, would you give it to your employees?" I got 100,000 committed and that took off.

Then I went to American Express and said, "What if we did a mailing"—I was still in college, 20 years old—"to your card members?" At the time, it was not a card you used close to home. "We'll give them incredible deals done in an American Express way. They'll be very upscale mailings." Then I went to the furriers, jewelers, high-end clothing stores, and top restaurants to sell space in this mailing. That just crushed it for American Express, so they asked me to do ten more cities.

At that point, I dropped out of school. That was the beginning of what evolved into Digitas. At first it was a coupon company, but I would redefine it to became a direct marketing company, then a customer-based management company, then ultimately a digital marketing company that used direct marketing and customer base management skills.

My next big break came in 1982 when I met somebody from AT&T at a tradeshow. They were going to divest and break up the Bell Phone System, and he asked if I had any ideas to address that AT&T was going to be 40% to 50% more expensive than the Baby Bells, regulated from discounting their price for some time. I flew home that night and had the idea to create "AT&T Opportunity Calling," which was building on the original concept of the coupon books.

"Since we can't discount the phone bill, what if we said that we'll give it for free?" The idea was that every dollar a customer spent on AT&T, we'd give them a dollar of savings on purchases, like American Airlines seats, GE appliances, GM cars, etc.? I said, "You have the greatest database in America. You know what people spend and where they call. You have their addresses and a long-term relationship." They committed $100 million to it, and I set out to pitch these companies. It was credited with boosting AT&T's stock price since Wall Street did not expect AT&T to have a marketing program. It became a long-term, successful program for them.

DP: These programs weren't just about getting new customers. They were also about providing additional value and building relationships that increased loyalty and retention. Can you tell us a bit about this kind of holistic view or philosophy of marketing and the customer relationship? How do you put together a program like that?

MB: We called our philosophy "behavior optimization." If a customer was exhibiting certain behaviors and you want them to do something different, like stay longer or spend more, then you have to think: What is that behavior worth? And, what are the barriers to getting the behavior? We would organize all of our ideation around levers. We'd look with honesty at whether the idea really would overcome the barrier.

It didn't seem very complicated. We built program after program upon the process of behavior optimization. We got our clients to do a lot of research and we broke everything down systematically. It was a combination of big ideas and the day-to-day customer management—through the phone, through the mail, and then ultimately through digital.

DP: What you describe in terms of levers and behavior optimization aligns with what today marketers refer to as the customer journey. Instead of thinking about getting eyeballs on ads or driving traffic to a website, you take the customer on a journey with deeper levels of engagement in each point of contact, thinking holistically about the brand experience. It's incredible that you pioneered this 30 years ago.

You were just 20 years old when you started working with American Express and 24 years old when you launched Opportunity Calling for AT&T, which at the time was the largest direct marketing campaign in history. You founded your latest company, UnReal, with your two teenage sons, Nicky and Kristopher.

I think a lot of especially young entrepreneurs shy away from founding businesses that service big companies because they either don't know how to pitch or land them as a client, or they somehow feel intimidated by them. At the same time, I think many big companies are struggling to serve the needs and connect with young people that really want to make a difference. This is one of the big themes of the book—how companies and the crowds can work together. What advice would you give to young entrepreneurs that want to work with big companies?

MB: That's a great question. There is an imaginary fear that big companies don't want to work with young entrepreneurs. I think it's absolutely not true. The challenge is how to break in. When I started working with American Express, I went to their local travel office and I just built a relationship with the manager. He said,

"I know somebody in New York," and then connected me to a woman that was a low-level marketing manager. She was able to make the decision. She eventually ended up becoming the CMO of IBM, which also became a client.

There is an imaginary fear that big companies don't want to work with young entrepreneurs. I think it's absolutely not true

Big companies crave creativity and innovation. I still maintain relationships with executives at my former clients, and I bring them young people who are doing cool things regularly. They're not part of the big agencies or technology companies or consulting firms. They're people with ideas. That's what they want. I think we forget that ideas are what make things happen for these big companies. The bigger the company, the less likely they are to consistently be able to deliver big ideas. They're craving them—entrepreneurs need confidence and to be fearless.

DP: That's actually really helpful because the next question I was going to ask was if you had a guiding philosophy or approach to scale from these kind of experimental days when you were just figuring things out until you grew to a thousand employees or more.

MB We only worked with big companies. When I led Digitas, we never had more than nine clients, and that took us to a thousand people. Everybody used to say, "You're crazy. You're at such risk. What if you lose a client?" I always said, "Look at the agency model. It's all about churn. At a 1,000 people, an agency our size would have about 30–50 clients and they'd be turning 10% to 15% of them constantly. They're using a lot of their best resources to pitch and that's why they lose their clients—because they're so distracted."

For me, I always loved our clients. What I did wasn't about the brand American Express. It was about the people that I worked for, and I'd do anything for them. I think that came across. Concentration is the opposite of risk. I think it's low cost to serve, to be deeply entrenched, and high cost to pitch. Yes, you have to be obsessed with doing a great job, or you will get fired, but you can build amazing companies on concentration, attention to detail, and deep relationships.

DP: Digitas was eventually acquired by Publicis. As you approached 40 years old, you decided to focus on helping families save for college. Many people might think that you would do that through founding a nonprofit or a scholarship fund, but instead, you started Upromise, which was a for-profit business that helped families save through earning rebates and rewards on purchases.

Upromise has been praised for being a pioneer in social entrepreneurship and creating an innovative financial business model that leveraged the power of the crowd and corporate partnerships to help over 10 million families save for college. And you did this many years before the term "collaborative economy" existed. Can you tell us about Upromise?

MB: I started Upromise because of my own situation with not having family savings to pay for college. That is what drove me to start the coupon books—to pay for expenses at school. So I wanted to help families get started saving earlier and save more, and I felt that you can't preach to them to save.

All of these college savings companies were out there saying, "Save money in a qualified 529 plan" and only a small percentage of families that should save, actually enrolled. Of course they want to save, but it's really tough. The hardest part is just getting an account open. What happens it that people say, "I'm saving for my kids' college as part of my savings account," but as soon as there is need for a new washer-dryer or a new car, they pull that money.

Our idea was to say, "We'll give you free money for college. All you have to do is open a free college savings plan." We got the biggest brands in the country to agree to give between 2% and 10% of what a customer spent into a college savings plan for their son, daughter, niece, nephew, grandchild—it didn't matter who, and accounts could all be linked, so lots of people could save for one or two kids. That was really cool, it made saving for college this big family affair where everyone could help.

The real thing that was happening that I don't think people ever understood about Upromise is that when you signed up for a free college savings account, you were signing up for a qualified account. You were giving us all of the information that we needed to open that account, so we had your kids' birth date, social security number, etc. We also had all of your credit card and store loyalty card info to track and automatically credit your free college saving.

Once we had this, we were able to go to you and say, "David, you have a one-year-old and a two-and-a-half-year-old. Did you know that if you save $22 per month of your own money between now and the time they graduate, based on some assumptions, here's how much you would have?" The point was that we already had you in motion. You didn't have to open an account because you had to put money in. You opened an account because you wanted to get free money for your kid or someone else's kid.

Over 10 million families signed up. We gave away over a billion dollars of free money, but we also got over $70 billion of people's own savings into these accounts. That was the idea. We made it fun and a collaborative group effort to help put kids through college.

It was originally conceived as a not-for-profit, but I quickly realized that to get the talent that we needed to build all of the back-end infrastructure it was necessary to make it a for-profit. My shares went into a foundation and the rest were made available to investors and employees. That was the idea and it was an amazing business model. Sally Mae acquired it in 2006. I wish I hadn't sold it. We could have taken that same idea into retirement or healthcare. We could have engaged the country to save for a range of milestones.

DP: Upromise is a flagship example of a company that serves the crowd—helping 10 million families put kids through college is awesome. You founded it in 2000, right in the wake of the dotcom crash and years before the concept of the collaborative economy, and back in the 1980s founded revolutionary rewards programs that were among the first to take a 360 holistic view of customer experience, retention, and loyalty.

There is this common thread throughout your career of building successful businesses by bringing partners together and creating new kinds of business models that are mutually beneficial for everyone involved. I think a lot of CEOs at big companies are struggling with this. On the one hand, they have the mandate to build shareholder value, while on the other hand, they're trying to do the right thing in terms of making the world better. How do you lead a company that's profitable and also benefits the world? What is the best way to approach leadership?

MB: I think the biggest issue and the biggest opportunity is consciousness. And I say that having personally worked with some of the biggest companies in the world and spending lots of time with their CEOs and leadership teams. A lot of leaders are caught up in optimization of business models rather than being conscious and looking at all of the constituents—including the planet, employees, customers.

> *Companies would be incredibly more successful with their customers and employees if they could collaborate with each other and solve needs from a place of consciousness*

I put planet in there very purposely. If most CEOs consciously thought about the planet, consciousness, and the whole crowd, they could solve for authenticity with their brand in ways that they're not thinking about. They're turning to the ad agency to solve for the brand, and they're turning to the employees with incentives for productivity, but what about, "We as a company, we're doing the right thing all the time, every day, authentically"? These companies would be incredibly more successful with their customers and employees if they could collaborate with each other and solve needs from a place of consciousness.

I had this dream after I was done with Upromise to set up a physical center that would bring together the most enlightened, aware, conscious leaders—I'm talking conscious gurus—and figure out how to bring through corporate leaders so that they can have this shift and see things differently. Some of these CEOs are still very close friends. Their lives are so overwhelmed. They have no time. They're sitting there on phone calls and meetings, under stress and pressure. They are missing the one thing that could bring purpose to their life and to the lives of every employee and customer that wants to do business with them. It's the ultimate collaboration from my perspective.

DP: On a flipside of that, several times when we worked together, you said, "The essence of good strategy is often what you don't do." This really stands out for me because I think today there's this sense that brands always need a new strategy for every new app or new platform or new social media innovation. They end up getting totally fragmented and lose focus. I have a keynote talk around this called "Don't Outsource Who You Are." Can you tell us about simplicity and focus, and the idea that strategy is often about what you don't do?

MB: I've seen it over and over. The essence of good strategy is sacrifice. It's being so clear about who you serve and how you serve them from a place of trust and collaboration, that you can get rid of so much extra stuff. So many companies are always about the next new service or product. It gets to the point where nobody can even keep track of all the stuff that they offer.

The essence of good strategy is sacrifice

What's your real promise? What do you do in my life? Not just for me but for everybody? And not the "We contribute *x* percent to the whatever-it-is charity fund." It doesn't matter. That's not real. I know it's marketing trying to influence me. Instead, *what do you really do that matters in the world?* Why can't companies start there? Why can't they convene their senior management with that question?

Focus on your purpose. "What do we do to create good in the world?" Not "What do we have to append to what we do?" Not "What can we market?" Get rid of all the other stuff. Just be that. Be the true value creator for the world and for your customers' lives. And, it's bigger than just your customers. Your customers are part of something bigger. Why are we not serving that bigger purpose?

It would be awesome if every company did that and used capitalism as a force for change for good. I think the potential of for-profit companies to create good is bigger than not-for-profit companies, by far. But that's not where they're coming from. The crazy thing is that's what customers and the market want them to do. They would improve their earnings, attract top talent, and be widely successful if they operated from a place of authenticity and purpose.

DP: You're applying a lot of these ideas around purpose and authenticity to your latest company, UnReal, in terms of authenticity around ingredients—where they're sourced, non-GMO and Fair Trade and these types of things. You also founded UnReal with your two teenage sons, and their authenticity and purpose comes through in the company's mission to "un-junk the world." Can you tell us about UnReal and why authenticity and purpose are so important for companies today?

MB: I'll tell you a story that illustrates this. I fell in love with this restaurant recently because it was one of the coolest places I've ever been. When you go in, it's clearly a values-based place. It's super hip. All the food is amazing. When I started talking to the owner, he told me where everything came from—everything is organic, the meats are grass-fed and more. I said, "You don't say 'organic' on anything. You don't say everything is from a local farm or the meats are grass-fed." He simply said, "You either are or you aren't."

He talked about his values, which communicated his purpose. He didn't have to explain authenticity because he was authentic. He shook my hand and I thought, "wow, that's really cool." He went on to say that he thinks it's annoying when you see labels like "organic," "local," and "grass-fed." This place has a line that's 25 people deep nonstop. This is on Cape Cod. I think to myself, "It's a year old. How does a place like this just become overwhelmed with people and they're constantly selling out of stuff?" It's because it's so damn authentic and purpose-driven. I think that's very much the future.

UnReal was founded by a 13-year-old who asked a question, in anger because candy, soda, and cookies were not allowed, "Why do the foods we all love so much have to be so bad for us?" And he was challenged by me with: "Well, maybe they don't..." We started the company with his stated mission to "un-junk the world." What we mean by that is, we want to create a ripple that creates change, by getting people to ask questions about their food. If we can make much healthier products for people and the planet that taste as good as America's favorites, people would ask themselves why can't all products be like this? We try to be as authentic as we can—sourcing the ingredients sustainably and doing no harm, trying to prove that it can be done the way we know it should be done.

> *We try to be as authentic as we can...to prove that it can be done the way we know it should be done*

Frankly, if the big companies truly cared, they could do it ten times better than we could. It comes back to consciousness. If they really cared about people's health, the farmers, and the planet, they wouldn't be using the ingredients and practices that they are and they could find a way to make it all work. Imagine if the CEOs of Hershey, Mars, Nestle, or one of these big companies sat down and said, "We're going to spend the next two years figuring out how we can do it. If we have to raise prices, that's okay, but we're going to do the right thing in everything that we do."

Wouldn't that CEO live a much happier life? They're smart enough and have the resources to be able to do that. They're not bad people; they're just not thinking that way. They're not actually asking that question and challenge their organization. All companies in every industry should operate from a perspective of "How can we leave the world a better place?" What a different world it would be. Given it's the only world we have, and the thing that matters most is love and relationships, how can this not be happening?

DP: One of the things that always impressed me about working with you is that you're a very giving and a compassionate person. You also have an incredible attention to detail and a lot of these intangible qualities that make you a great leader by inspiring greatness in others. These things seem to be part of who you are, and it makes me wonder how much leadership can be taught.

You dropped out of school before finishing your degree, but then you also started an entrepreneurship program and you were on the board at your college, Boston University. You also started Upromise to help families save for college. Clearly you believe in the power and importance of education. Given all your experience, can you teach entrepreneurship and leadership? If so, how do you do that?

MB: First, thank you for such a special compliment. I think you can. It depends upon the student. The reason is, I think that the best leaders and entrepreneurs tap into authentic passion and purpose—the purpose to do something great and the passion to be your best self; to fulfill your true potential. Not everyone has these qualities and they aren't qualities that you can teach. They come from within and motivate everything that you do.

People with passion and purpose will usually seek out advice, guidance, and help. You can be inspired by others—that's why I think the best teaching is case studies and learning how people have done things before. It's important to hear their stories and that can lead to learning what that you don't know how to do.

There are also the entrepreneurs that do things without advice and direction. They just go do it, work their way through the nuances, and figure it out. Like when I saw that coupon book, I had an idea, and I just did it. But the truth is since I was in third grade, I was always doing something—I grew up in Miami, so I was cracking coconuts, washing cars, whatever. I certainly didn't think that I was becoming an entrepreneur. I just did each of those things from a place of passion and purpose. When I started in business, I had no clue how to do accounting or "business stuff" but I tell people I didn't *need* to go to college because I just needed addition, subtraction, and multiplication. It was largely instinctual and I sought examples and people to show me. If you have that inner drive—that passion and purpose—you're going to go learn what you need one way or another.

If you are purpose and passion-driven, you can be a great leader because leadership is about treating people the way you would want to be treated, being fair, and realizing that when you do that, it comes back. That's not *why* you do it (to come back), but it's true. When you go through tough times, don't write good people off; find ways for everybody to stay together. There is a personal reward in doing the right thing that people underestimate.

I think if you're purpose and passion–driven, you can absolutely learn everything yourself, through a combination of inspiration and education. Yes, you can teach entrepreneurship and leadership, but inspiration may be more important than education—seeing what other people have done and believing that you can go for something that really matters and excites you. To borrow a line from Dave Quirke: Believe that everything is possible. That is the defining line between an entrepreneur and a non-entrepreneur. If you believe it can be so, it *can* be so. If you don't believe it, it cannot be so.

Believe that everything is possible. That is the defining line between an entrepreneur and a non-entrepreneur

DP: Imagine that this whole interview has been in front of a classroom of aspiring entrepreneurs, As a final question: What parting words of wisdom would you give them?

MB: In anything that you want to do, no matter what kind of company you want to build, you can add a component to it that makes a difference in the world and has a positive impact on the lives of others. A lot of people ask me: "Would you take a look at what I'm doing? Would you consider investing?" I first say, "I'm not going to invest in anything that isn't going to make the world a better place."

Anybody that wants to change the world and do something good, with a good plan for how to do it, is going to attract good energy and support. There's no question in my mind. That's also what most investors want to get behind. They look for entrepreneurs that want to do something good and have a good economic business model. Everything else, to me, is a snore. Don't do anything just to make money. Create a better world at the same time. Lead from a place of purpose, passion, and love.

....

MICHAEL BRONNER is the founder and chairman emeritus of Digitas, a global digital marketing agency employing over 6,000 people. Digitas was purchased by Publicis Groupe in 2006. Michael also founded Upromise to help make college more attainable for middle class families. Upromise is the largest private source of college funding, with over $1B of free money given to participating families. Upromise was acquired by Sallie Mae in 2006. In 2012, Michael co-founded with his two sons, UnReal, a new generation food company, with a mission to "unjunk the world" by reinventing America's favorite junk foods, making them without the bad stuff.

....

ARENAS NOT MARKETS

Rita Gunther McGrath

Rita Gunther McGrath is the top business strategist in the world. *The End of Competitive Advantage* shows that much of what we believe about strategy is based on assumptions that are no longer true. The world is now highly interconnected and constantly changing. Competition can come from almost anywhere. We compete in arenas, not marketplaces. We need a new strategy toolbox.

- Twisting markets to compete in new ways
- How the Internet accelerated changes in strategy
- Companies can't shrink their way to greatness
- The future of work and sharing prosperity

Large organizations have scale and resources, but often struggle to adapt and innovate. Her pioneering approach to strategy integrates frameworks familiar to startups and entrepreneurs like customer development and business model innovation from Steve Blank and Alex Osterwalder. This versatility makes Rita an expert in navigating paths to success in rapidly changing and volatile environments.

DP: The preface of your book The End of Competitive Advantage *opens with the sentence "Strategy is stuck." This has to do with legacy assumptions about competitive advantage that you think are wrong now. The first is that industry matters most; and second, once achieved, advantages are sustainable.*

In contrast to thinking about competing in marketplaces shaped by industries, you propose strategies focused on arenas or spaces that are not industry or market specific. Can you tell us how strategy got stuck around the notion of competitive advantage, and explain your new strategy framework?

RM: The strategy field as we know it took hold in the 1970s and 1980s with the flowering of tools like Five Forces Analysis, the BCG matrix, tools for portfolio analysis, the value chain, and those kinds of ideas. That material became embedded in business textbooks and how we teach and think. What these tools all have in common is an underpinning alignment to this notion of stability and sustainable competitive advantage.

When you think about the era in which those tools became normal practice, it was very different than today. India and China were closed. South America and Africa weren't doing much. Mostly American and European firms had a lot of markets pretty much to themselves. There was competition, but it wasn't completely global, digital, fast-moving competition. The world changed, and strategy as a field became stuck.

We really need a whole new way of thinking about strategy...a whole new strategy playbook

One of the purposes of my book was to unblock that a little bit and say, "We really need a whole new way of thinking about strategy. We need a whole new strategy playbook." Although my book is called *The End of Competitive Advantage*, I'm beginning to be of the mind that the very term "competitive advantage" almost leads us astray, because it says you should be focused on the competitors, who by definition are other companies in your industry that compete to do various things.

The arena concept has leadership defining the space in which to compete, and the space may not have anything to do with conventional industry boundaries. For example, think of the kinds of definitions of space when a technology company like Google goes after advertising, which traditionally was the domain of the media industry. What they did was say, "We can get users on our site looking for things, and what better way to pitch ads than to users on our site looking for things."

What you're seeing are companies saying, "Hey, we can twist this market and look at it in a really different way." When you think about a lot of the high-growth firms over the last few years, what they've basically done is that. They've taken an existing set of market relationships and twisted them—the hot startups like Uber and even more traditional companies like Ikea have some kind of edge with their customers and twist convention.

DP: The pillars of traditional strategy are based on assumptions of a more stable, predictable world that existed before the Internet and even before widespread international trade. Big companies need a new strategy tool set to integrate the constant change and unpredictability that arises with the digital world.

The digital world, in turn, is being layered on top of the physical to facilitate more efficient exchanges of goods and services through the creation of new types of crowd companies. The collaborative and sharing models from Web 2.0 and social media are now empowering people to get what they want from each other without necessarily needing brands. These are the types of changes covered in this book.

How has the digital world and the collaborative economy impacted the changes you see in strategy?

RM: I think they accelerated the changes. One thing that is different is the Internet makes it much easier to build a platform that others can tap into. Before these Internet businesses, the transaction costs were too high or the difficulty of getting the information was very substantial. Take McKinsey as an example. If you wanted to figure out where to put a bank branch in 1980, what did you do? You hired a bunch of McKinsey consultants with stopwatches and clipboards that stood on street corners and counted how many people went around. Today, you can get that same information from a Google search.

The theory that I've been working on has to do with markets vs. hierarchies, which relates to the collaborative economy. Basically, the idea is hierarchies are necessary under certain conditions, such as when the quality of the information is opaque (neither party to a transaction is able to establish a realistic value), or when there are strong incentives on the part of either party to cheat. There are certain conditions when transactions are expensive. These types of transactions lend themselves to being conducted within an organization.

Hierarchies are necessary under certain conditions

I think the digitization of the economy made it possible for markets to emerge where before you could only have organizations. When you think about things like eBay's user rating system, Amazon's comment system that can inform prospective buyers of the experience of previous buyers, or Uber drivers and customers being able to rate each other, those are all examples of how markets are creating the kind of trust that used to be the job of organizations to provide. Now they are being replaced by collaborative kinds of markets.

On top of that, you have the power of network effects, which means every single additional user adds to the value other users get from participating in these markets. Once you have accelerated adoption rates, it creates the potential for exponential growth. Ironically, that can become a sustainable competitive advantage, because once you're the marketplace where everybody wants to be, for whatever reason, that makes you stronger and stronger. It's hard to overcome that sheer volume of market exchanges. In this context, I think of the collaborative economy not necessarily in an intentionally collaborative way, but more in terms of transparent market transactions.

Then the question is: Will there come a tipping point at which that begins to unravel? Once your grandmother is on Facebook, is that really where you want to be? If you look at the social media sites that have failed (such as MySpace), you see that the network dynamics act in reverse—as users start to leave, the network becomes less and less valuable to those that remain.

DP: My last book Disruption Revolution *looked at the revolution of innovation that occurred in the aftermath of the 2008/2009 economic crash. There was a basic idea that Bryan Solis talked about that constraint drives innovation. Cutting back on resources forced companies to find new ways of generating revenue. Meanwhile, entrepreneurs rallied around the term "disruption" to create all of these new innovative products and services. This became a catalyst for the collaborative and sharing economy with the meteoric rise of companies like Airbnb, Uber, Lyft, TaskRabbit, and so on.*

So much discussion about innovation and economic recovery focuses on disruptive startups. How do the big companies that you work with view the trends since the crash? What can we learn from what happened, and how might they approach strategy differently going into the future?

RM: Let's start with your idea that constraint drives innovation. Any design thinker can tell you that what designers want most when they start a project is to understand the constraints, because that helps drive where the creativity is going to be focused.

Post-2008, the basic strategy was to stop everything, downsize, cut, get rid of anything unnecessary, and try to right the ship. The downside of that was a lot of people working on innovation either got booted out or were taught to funnel their innovation towards cost cuts. Then starting in around 2010/2011, we started to see more of these companies beginning to say, "Hey, we need to get back to growth."

We need to get back to some kind of innovation, because you can't shrink your way to greatness. Lots of companies struggle doing both—running today's business and having the right practices for a system of innovation—so there's a lot of talk about innovation, but not very much actual innovating.

The public markets exacerbate this because you can't make a plan to drive innovation that diminishes the pressure in the short term to forecast earnings. Your typical manager will say, "I can get by for two, four, six, eight more quarters before lack of investment in the future catches up. What's the incentive?"

We need to get back to some kind of innovation, because you can't shrink your way to greatness

DP: I love your comment that you can't shrink your way to greatness. It's a simple statement of the complex problem that so many companies face being trapped in the cycles of quarterly earnings reports and struggling to make the necessary investments in innovation needed for growth.

If the world is constantly changing faster and faster, leading companies to compete in arenas instead of marketplaces, how should they think about advantage and approach innovation?

RM: I think of advantages in terms of waves. You've got the beginning part of the wave where you can see the differentiating advantage, incubate it, get it into the market, and ramp up. Speed is critical here. If you ramp up too slowly, then competitors can match your efforts.

Then you have the exploitation phase of the wave. It's a steady state. You're doing great stuff for today's customers. Then if the advantage erodes, you have the disengagement process that involves reconfiguring your team, assets, and capabilities to transition to another advantage. The disengagement part is often done very painfully, with a lot of damage and not much value recaptured.

I would say today's large organizations are pretty good at running the exploitation phase. I think one of the reasons my book has been so well received is most companies know that they can't continue to exploit, exploit, exploit, but they don't know what other tools to put into their toolkit. People like Alexander Osterwalder are trying to crack that. What are the tools to help companies manage these complete cycles?

We need companies to build innovation as a proficiency. Most companies think their problem is getting good ideas, so they do ideation boot camps. But ideation is only the first part of innovation. After you get the ideas, you need to incubate and turn them into a business concept.

Build innovation as a proficiency. Most companies think their problem is getting good ideas...but ideation is only the first part of innovation

One of the big shifts in innovation is the increasing supremacy of design thinking. People are embracing those principles and saying, "Yes. Let's build prototypes. Let's do customer development. Let's make what Steve Blank calls 'minimum viable products.'"

DP: A lot of strategy traditionally focused on competitive advantage in industries, and they also used words like "competition," "exploit," and "battling" to describe this cutthroat state of the marketplace.

In this new framework of transient advantage that you present, what role might collaboration or sharing play, such as forming strategic partnerships or leveraging relationships with fans?

RM: The first thing is that there are a tremendous number of large organizations today that simultaneously compete and cooperate. Samsung and Apple would be examples. One part of the organizations competes like hell and then other parts rely on collaboration.

Part of what we're seeing is a breakdown of the traditional notion of what a competitor is. There are a lot more tactical decisions made to cooperate for whatever reason, like common enemies or "the market's moving really fast and those are the guys that happen to have the solution at this particular moment, so let's collaborate with them."

In terms of collaborating around customers, I have a concept that I call "the consumption chain," which basically says any potential customer is constantly meshed in these sets of experiences—developing awareness, searching for solutions, finding a solution, picking a provider, making payments—that are a whole chain. Most companies today have a pretty primitive understanding of how customers make their way through that whole chain, so they focus on the pieces of it that they sell or that they work on. They don't look at the whole experience. When you think of the value of collaborative relationships, a lot of times what that means is to fill out the whole chain and make it a complete experience. If the chain's broken, the experience breaks down and the customers may go to someone else.

A firm that does a good job of this is Amazon. If you think about Amazon's business model, they have something at every single link in the chain. For example, search. How do I become aware I have a need? "Well, people who have bought this book also bought that book." "We noticed you bought titles by so-and-so. Would you like to be informed when their next book comes out?" "We've got recommendations for you based on profiles of people similar." That one link in the chain has fifty possible points of differentiation. The driver of collaboration is the need to get into customers' experiences in a way that companies never had to before.

Conversations between customers and companies are also changing. This involves the whole idea of social marketing, having back and forth communications with your customers, creating authentic messages, etc. This has become high on the agenda of senior leaders because it's a two-way conversation now and companies don't control the message. The message is out there, and they have to figure out how to plug into it authentically or they'll be at a disadvantage.

The conversations can have a spiraling effect. Once a cadre of people gets upset enough, you can have a real problem as a company because the message out there may not be the one you want to see. I think for customers it's liberating to be able to ask 12 people in the network that you trust, "What do you think of the service? What do you think of that product? Which phone operator are you happiest with and why?" The questions and knowledge exchanges opening up all over the place are very empowering.

An entrepreneur that I interviewed mentioned the social media version of a very traditional technique, which is figure out where your customers' pain points are. He will troll through the comments made by customers or users and look for certain words—"I love," "I hate," "I wish," "wouldn't it be great if"—and then see if he could build a business around solving that pain point. Often, social media provides the validation for an idea that entrepreneurs come up with in solving their own pain points. Social media management service Buffer got started this way. Its founder wished for a way to enter tweets but have them spaced out throughout the day and used social media feedback to refine the idea before launching it in a very simple, bare-bones way.

DP: When Etsy went public, in their IPO, they made 5% of their stock available to store owners that built businesses on top of their platform. This promotes loyalty and community in a critical growth phase by basically making their best customers into strategic partners with a vested interest to grow the business.

How might companies approach relationships with their shareholders, investors, or most loyal customers differently based upon the implications of your ideas?

RM: I think right now we are in the grip of what Steve Denning calls the worst idea in the world, which is that companies exist primarily for shareholders. The average shareholder in a large public company in America holds their shares for seven days. Running your entire organization for that is a bit crazy.

I think we need a balance between various stakeholder interests. We had a much better sense of that balance years ago, as the work of William Lazonick has pointed out. He wrote a fascinating article in the *Harvard Business Review* called "Profits Without Prosperity." He was able to show that until about the mid-1980s, corporations took their profits and invested some for the employees, some for their returns to investors, and some for investments in the future.

In the 1980s, three things happened. First, Congress changed the rules on stock buybacks and made it much easier to buy back your own stock, which has the effect of raising the stock price. Second, was the ascendance of the shareholder returns concept.

Third, as an unintended consequence of laws that tried to limit CEO compensation, it became common for CEO compensation to come from stock price appreciation, stock options, and stuff like that. In fact, a recent study featured in the *New York Times* clearly showed that companies that over-use buybacks suffer long-term slowdowns in growth compared to similar companies that were less prone to giving buybacks and dividends.

What Lazonick argues is that you can see how, up until about the 1980s, the pace of worker compensation, investment, and productivity improvement kept pace with growth. Then you saw returns to labor trail off, even as productivity continued to rise. We need to rethink the relationship between the owners of capital and the owners of the means of production and labor.

We have to go back to a very traditional idea in economics. Because right now, the owners of capital are being treated as though capital is a scarce resource that should be conserved, when it isn't. There's money everywhere. People are desperately chasing returns right now, which is causing asset bubbles all over the world.

People are desperately chasing returns right now, which is causing asset bubbles all over the world

What we're seeing is this kind of bifurcation (some people call it the "Hourglass Economy") where those with rare skills, or those who have a particular network, are getting compensated very well; those at the bottom are not doing well at all; and in the middle, it's just being hollowed out. And all the growth is coming at the very top or the very bottom. I think we're ready for a real rethink of the relationships between stakeholders and investors, how companies invest, and what they invest in innovation.

My research suggests if they don't get that balance right, they might succeed quarter by quarter in the short term, but they won't last. The other thing that has really changed is the ability to do things very quickly at scale. The small entrepreneurial firm that wants to do something can make it happen very fast. Using market-based services such as Amazon Web Services and contractors from companies such as Upwork allows even small companies to get to scale quickly and move fast. Big firms with budgeting restrictions and lots of politics are much slower.

DP: You work with and study large organizations with billions in revenue. In the last few years, there have been many billion-dollar companies created from the tech space like Facebook, Twitter, Airbnb, Uber, and so on. How do large companies remain competitive with tech companies entering these arenas, and how does the massive reach and scale of these new platforms impact strategy?

RM: The first observation I would make is the big companies never worried much about Silicon Valley because they were kind of doing their own thing in different industries such as semi-conductors and communication. It didn't bother Nestlé or the taxi business or the hotel industry. The most recent wave of Silicon Valley startups is gunning for the establishment, so there's a lot more nervousness in management among the big companies. That being said, big companies have a lot to work with.

One of the firms I'm a faculty partner with right now is a company called mach49, which is an incubator designed to help global Fortune 1000 companies create new businesses from within their organization. Their thesis is big companies have assets, talent, and resources that entrepreneurs can only dream of, but what they don't have is the way of getting those to market fast in a non-bureaucratic kind of way.

I think you'll start to see big companies using incubators, accelerators, and that kind of thing a lot more than they have in the past. With luck, you will see big companies learning to control the power dynamics and the politics that often get in the way with respect to innovation.

DP: We covered lots of ground around strategy and big companies. There is great speculation about the future of the workplace, including the threat of automation and AI described by Martin Ford, author of The Rise of the Robots. *As a final question, I'd love to hear your thoughts on how big companies should approach hiring differently so they can succeed in this new era of collaboration, sharing, and disruption? What does the future of work look like to you?*

RM: Younger generations coming into organizations are bringing a lot of fresh insights, perspectives, and unencumbered thinking. I think that will produce some really surprising outcomes. They relate differently to people at work and think differently about what work is. It's a thing you do, not a place you go to. We're going to see a real transformation of the workplace as we know it as this group—which is now the largest group of employed people— begins to ascend to more significant positions of decision-making authority.

Data is also changing the way that we make decisions. There was just an article in *The New York Times* about how a lot of these younger Internet companies don't need middle managers to make decisions because they run an A/B test and automate that. You have different ways of dealing with vast amounts of data corralled so that you're not making guesses, judgments, and hypotheses. You're actually able to say, "Let's test these out and see what the right answer should be."

We are increasingly going into what Reid Hoffman calls a "tour of duty economy." If you think about that in terms of human resource practices, you might bring somebody on to do innovation work, and when that tour is up that person either renegotiates for the next gig or they move on. Maybe they come back again when there's another opening for their skills, or maybe not.

We are looking at dynamism in hiring practices because we're not doing career ladders with fixed mandates. We build jobs around people in the context of a tour of duty. There are already firms that operate this way like Accenture or the big building construction firms. They bring together the right skills when something needs to be accomplished, and then those skills go somewhere else.

The way that we structure organizations is going to change. When I'm optimistic, I think the workplace of the future is going to be great in many ways. People like you and me will be able to have periods of time with a company and periods of time doing something really different. We're going to be able to shape these interesting careers in a much richer way than people were able to in the past.

When I'm on the pessimistic side, I worry that organizations and people that have not figured out this transient advantage stuff will start to treat people like disposable cogs. We see this in services and fast food. The attitude among the hiring managers is "Ugh. They're completely replaceable. Why pay them more than minimum wage?" and I think that's terrible.

I'm hopeful that we will start to recognize this is not the kind of society we want to build and it will start to shift. I'm doing some work with MIT and one of the things we are interested in is why more companies haven't figured out that if you use human beings to increase revenues, that's actually more valuable than taking the same human beings and treating them as a cost—because there are a lot of things human beings can do to drive more sales, create more expertise, and connect and build relationships than people that are just being treated like a pair of hands.

We have a choice here about what kind of society we want to build and how broadly prosperity is shared

We have a choice here about what kind of society we want to build and how broadly prosperity is shared. Without sounding too Utopian, new technologies show promise to give us healthier lives, more flexible career structures, and more interesting jobs. To make that happen, however, we need to re-write some of the rules of competition and career structures. At the better companies, this is already happening.

....

RITA GUNTHER MCGRATH is a globally recognized expert on strategy, innovation, and growth with an emphasis on corporate entrepreneurship. Her work and ideas help CEOs and senior executives chart a pathway to success in today's rapidly changing and volatile environments. McGrath is highly valued for her rare ability to connect research to business problems. *Thinkers50* named Rita in their top ten global list of management thinkers overall for 2015 and 2013 and presented Rita with the #1 award for Strategy, the Distinguished Achievement Award, in 2013. Rita has also been inducted into the Strategic Management Society "Fellows" in recognition of her impact on the field. She consistently appears in lists of the top professors to follow on Twitter. McGrath is the author of four books; the most recent being the best-selling *The End of Competitive Advantage* (Harvard Business Review Press), rated the #1 book of the year by Strategy+Business.

BETTER BRAND VALUES

Alex Bogusky

Alex Bogusky left advertising shortly after being named Creative Director of the Decade 2000–2010, the industry's most prestigious award. Under his leadership, Crispin Porter + Bogusky pioneered modern brand building and viral marketing. Now he does meaningful work with people he loves, from helping Al Gore with climate change to being an advisor and investor in startups like Lyft and HealthySkoop.

- If you have a business, you have a brand
- The rights we demand vs. the rights we deserve
- Ask questions even if we cannot answer them
- How bullying holds back the evolution of society

Alex is one of my personal heroes. He has uncompromising ethics and convictions, is bold enough to speak out and push boundaries around issues that matter, and had the courage to walk away from an incredibly lucrative position to do work that positively impacts society. This is a thought-provoking, wide-ranging interview with one of the world's most brilliant creative minds.

DP: *We both left advertising out of a need to do more with our lives. I was Senior Director and account lead at the social media agency of record for Volkswagen, and CP+B was their creative agency, so we also shared the same client. I sold everything in New York City, bought a one-way ticket to Thailand, and spent six* *months in Southeast Asia and India. That story is the basis of my first book* Red Bull to Buddha.

One of the things that had a profound impact on me during my travels was the encounter with poverty. It made me rethink what's most important and led me to focus on how I can give back in my life and work. I thought it would be interesting to open the interview with that because you have an interesting hobby of collecting signs from homeless people. Can you tell us about that?

AB: I found the signs interesting because I saw them as this form of advertising that people pay attention to. Sometimes they're very creative, so as an ad and creative guy, I enjoyed what I saw. Plus I imagine I'm like a lot of people. I don't know what to do, like do I give? Do I not give? Is it right to give? Am I doing the wrong thing by giving because a lot of people tell you it's the wrong thing to give? Do you look forward? Do you look at them? Do you...? There's all that.

I thought it would be neat to buy the signs that I think are creative and that will be my way of giving. I value creativity so it will simplify that little aspect of my life. I don't know what to do with them, but I'll do that. Then I was like, Do I give them a marker and cardboard because I can't leave them without their sign? That's how they make a living.

Finally, I felt all set and roll up to this guy. "I'd love to buy your sign. It's very funny. It's very creative," I said. "Would you consider selling it?" and he was like, "Yeah. Totally." The first time, I didn't do it from the car. I parked and walked around because I felt like this could take a little while. He had a lot of fun with it, brought his buddy over, and they thought it was crazy that I was buying the sign.

I was a little worried, like, "Are they going to be offended? How are they going to take it?" And then I said, "Hey, I have cardboard and markers," and they were like, "No, no. We're good. We got cardboard and markers for more signs." No one has ever taken me up on the cardboard and the markers. That was just something I assumed would be important. And no one has ever been upset.

They get a kick out of it and they get excited so now I do it. It's really quick. I just roll up.

Sometimes I think people know there's a dude that buys signs in Boulder. Then, of course, friends hear that you buy signs and they'll buy you signs, so the collection's grown, but I haven't done anything with it. I just save them at this point. I've made some t-shirts out of some of my favorites and donated money from sales to the homeless shelter.

Mostly I learned about giving. You hear all the time that there are lots of reasons not to give, and I think everyone in our culture can repeat the reasons that you shouldn't give to someone who's begging or asking for money on the streets. You wind up talking to people a lot, finding out shelters are fine, but they are not great places to be. Stuff gets stolen, fights break out, people are afraid for their safety, and there are lots of contagions. I'm not saying shelters are bad, but they're not for everybody in every situation.

People also say they will use the money to buy drugs and alcohol. Well, I had a thousand people on my payroll and lots of them used the money to buy drugs and alcohol. No one ever said to me, "Don't pay them." If that's what they're going to do with it, then that's what they're going to do with it.

Right now, the best thing I can do is just say, "Hey, somebody cares enough and I've got a little extra, and it's good to give. Generosity is a good thing to cultivate." What I learned from the process is don't overthink it. If you want to be generous, just be generous.

DP: That's a good takeaway. Many readers of this book work in startups, or they are aspiring entrepreneurs that have little or no experience building a brand. I think it's safe to say you're one of the world's best at brand building. You transitioned from the agency world to become an investor, advisor, and founder of startups, including BoomTown, one of the leading accelerators based in Boulder.

Imagine you are speaking to a room full of entrepreneurs that may not understand branding. Why are brands important and where do you start building one?

AB: I have a very basic notion around branding, and in anything, I try not to use language or sophisticated terminology to distance ideas from people. If you have a business, you have a brand. And you can't avoid it. You have a brand. The question is really: Can you build a good brand, or is the brand going to be neutral or sort of bad?

If you have a business, you have a brand. The question is really: Can you build a good brand, or is the brand going to be neutral or sort of bad?

We often rename startups because they come in with terrible names. I don't think a name has to be great. The name and the brand are an accumulation of everything that you do—the communications, business, interactions—and those all fit to whatever is your brand. It's prudent when you're small and young to find a brand that suggests what you do. Also, don't get too attached.

Early-stage entrepreneurs often came up with their brand name before anything else. "The name of my product is Blah Blah." That's not necessarily their expertise. When someone is not flexible around the name, they also tend to not be approachable in any other way. One of the first things I'll say to test an entrepreneur is, "How would you feel about changing the brand name if you needed to? If it turned out to be necessary?" If they absolutely think it's a terrible idea without any consideration, that's more of an indicator that he or she is not a very good entrepreneur—at that early stage. It's one of the things that you can do that can have the biggest effect at that early stage. Later on, you can't necessarily do it.

I'm also excited about the march of design as an important element. I think tech lifted the profile of design in the last 10 years, and if you go back to business books 20 years ago, design doesn't come up very much. Every business book that deals with tech startups now seems to reinforce the notion that design is hypercritical. Lots of successful tech startups will have a designer on the founding team, which is pretty radical. As a guy that started as a designer, I kind of love that.

DP: Instagram is a good example of that. They were very design-conscious, and that was one of the things that really led them to differentiate. There were a million different photo apps at the time. So many, in fact, that it's become kind of an inside joke in Silicon Valley to talk about building a photo app, which is to say you don't have an original app idea to work on so let's try to jump on the bandwagon of what everyone else is doing.

AG: You'll see big VCs advise people, like, "Go out and find a great designer"—because there's more value in design, and I think people understand it. Maybe tech in some ways has understood it because its flexibility and real-time ability to change highlights the effect that good design can have. It's instantly apparent. Prior to the Internet age, you couldn't see the effects as quickly or at all. It would take six months to notice a design change I think vs. today, which takes six minutes when you get the instant feedback from data.

DP: When you left advertising, you partnered with your wife and Rob Schuham to start the Fearless Cottage. There was this idea of being fearless in terms of taking on the impossible, and also to fear less and love more, so the name had an interesting double meaning. Fearless evolved from an incubator of ideas and projects, into Fearless Unlimited, a social impact agency.

One of the Fearless projects that I found particularly interesting was a new Consumer Bill of Rights, which encouraged companies to be transparent about things like donations, ingredients, labor, etc. Can you tell us about Fearless and what led you to advocate a new Consumer Bill of Rights?

AB: I had so many large clients by the end of my ad career. You become a mouthpiece for every client, and so it's difficult to say anything and navigate every issue that your clients have. It was kind of a maze. My words would go through a maze to get out. Eventually, that maze became so dense that I felt like I couldn't say the things that I believed. I don't know if a lot of people wind up in that position. Some people would say that describes everyone in corporate America. I don't really know, but I found myself there.

I left the agency business because I wanted to get back to being able to voice what I thought. At the time, my thoughts were fairly heretical and upsetting to corporate America around things like transparency, ingredients, and GMOs. Those issues have gone mainstream, but six years ago were radical. During that arc, Fearless got created. It was a kind of experiment for exploring these ideas and doing work that we believed in with nonprofits and with Al Gore on climate change.

People will treat you the way that you allow them to treat you. I think that with consumers, it's the same contract: You will be treated the way you allow brands to treat you. When you ask, "What rights should Americans have?" If you ask the British in 1770, they had a very different answer to that than where we wound up. You have the rights that you demand; you don't have the rights that you deserve.

People will treat you the way that you allow them to treat you

Some consumers are very aware of that contract. They think about it, work on it, and press for it. But the vast majority don't care. Businesses are beholden to their customers. They're going to behave in whatever way customers want and they'll live up or down to whatever standards customers demand.

Some visionaries like Patagonia, Ben & Jerry's, or Paul Polman at Unilever will say, "No one's asking me to create a higher standard, but I'm going to go do it." But most companies stay barely in front of consumer expectations. That's the way the world works. People will have the brands and transparency that they demand, and they won't have anything more.

DP: You went from building some of the biggest brands in the world and being awarded Creative Director of the Decade from 2000–2010, to applying those same skill sets to nonprofits and climate change. It seems like now companies are trying to have a more positive impact on the world, while many nonprofits struggle to make a difference that really breaks through to public consciousness. Can you reflect on your experience building brands across the entire spectrum?

AB: Nonprofits are way more messed up than the for-profit space, and messed up in a way that would surprise most people. You think they would be more cooperative than for-profits, but it is absolutely the opposite. You see much more cooperation between for-profit companies than nonprofits. The reason is the way both are funded.

Big foundations and endowments have big pools of money and evaluate nonprofits based on their success. Not the audacity of what they took on or their incremental progress, but "We said we'd do this and we did it. And we told you last year we'd do this and we did it." As a result, nonprofits promise to hit minute little goals that measure success and are on a funding treadmill. They can't cooperate with other nonprofits because they all fight over the same pool of money, and they can't take on real issues because if they fail then funding will not get renewed.

We are not getting out of our nonprofits what I think most people want and expect

There are fewer and fewer examples of nonprofits like Greenpeace willing to take on impossible challenges like driving your boat between the harpoons and the whales, and chaining yourself to trees, or like 350.org taking on the Keystone Pipeline. When you think of the nonprofit space, you imagine an attitude of "It doesn't matter if we win; we have to try" but in practice most nonprofits only focus on things that they can win because of the way that fundraising works. We are not getting out of our nonprofits what I think most people want and expect.

The same type of thing happens in politics. Al Gore told me that before Citizens United, you would run, you would win, and then you go legislate. The heightened polarity of every issue would exist during that campaign. Then it would end and both sides of the aisle would work together to legislate. Eventually, you campaign again because the fight generates funding. Now, you need so much money that you can't stop fighting because without the fight, there's no funding. Sides don't come together because they can't afford to stop fighting and it's so expensive to run a campaign.

DP: At one extreme, we have nonprofits that don't take on the impossible because they are trapped into focusing on small achievable goals to keep getting funded, and at the other extreme politicians constantly focus on the impossible because the polarizing messaging and fight creates a constant state of emergency to raise more money.

What this suggests is that billions upon billions of dollars pour into efforts that may never make any meaningful, lasting difference. It makes me think of Peter Diamandis and the idea that the best way to become a billionaire is to solve a problem for a billion people. Look at what Elon Musk is doing with Tesla and Solar City to wean us off of fossil fuels and reduce our carbon footprint.

If nonprofits and politicians aren't getting the job done, should we expect the for-profit sector to change the world and take on the impossible? How could corporations work with the crowd to take on big problems?

AB: I think you see the private sector stepping in where government used to be, because government's going broke and they can't do it. The tax base has shifted. There used to be more corporate tax in the 1970s, and now the tax burden is more on individuals. Corporations need to step in where there's been that gap in funding where government used to be.

They're compelled to in two ways. First, they might be compelled to because they see a need and they might see an opportunity to do something that builds their brand. Second, they do it because no one in the system can afford for this thing to collapse. It's just plain necessary to step in. It's similar to the shift where a lot of social services were done by the church and they moved into government. Now they are moving out of government, and a lot of what are essentially social services will need to be funded by corporations working together with community. I don't have an idea of how or the best way to do it, but I identify it the same way you do. I see it as a megatrend.

DP: *This notion of identifying a problem, without necessarily knowing what the solution is, points to the importance of asking questions. In several of your talks, you emphasize the importance of asking questions even when it may not be technologically or economically possible to answer them.*

In your own work, this translated into excitement around pitch events and optimism that innovation can help solve the world's biggest problems even though they might seem impossible today. I share the same philosophy. This book in many ways is about asking impossible questions in an attempt to shape a larger conversation. Why is asking questions so important even if we can't know the answers?

AB: People don't ask questions because of intimidation. When you have a culture of fear it can be difficult to ask questions because you're going to be attacked. Look at Twitter five years ago. You could tweet or retweet. Today, the retweet hardly exists. There's a "like." You tweet something and then tons of people will like it because that's pretty safe. Retweeting says, "I believe this too" or "I believe other people should think about this," and that behavior is down on the platform.

There's a giant social cough around suggesting that you believe something because there is a risk of backlash. There is a culture of fear and intimidation that—I don't know if it's conscious—works to get people not to say things. When that dialogue doesn't take place, change can never take place. Dialogue in front of change really matters, and questions drive the dialogue. They never go in any other order. There is never change and then people talk or ask questions about change.

I might be willing to tweet about something that other people aren't, like climate change and GMOs. For somebody else, it might be a much smaller increment of what they're comfortable with. If everyone pushes a little bit from where they are to a place where they're a little bit uncomfortable and lean into that, that's doing your part. If that doesn't happen, then we really are fucked.

We have this interesting opportunity where social media is still the Wild West, but every day you see people working to get it under central control. There's a window for humanity to keep pushing and make sure that doesn't happen, because mass media has never been owned by the people until about 10 years ago.

Humanity now consumes more peer-to-peer media than top-down, state-down, corporate-down. That's a massive shift and lots of folks don't like it, which is why it's important for everybody to keep pushing whatever little bit you can and where you feel comfortable. Own and take advantage of it, because it may be a limited time offer.

> *Humanity now consumes more peer-to-peer media than top-down, state-down, corporate-down. That's a massive shift and lots of folks don't like it*

DP: *One of the things that comes up in your writing is this idea that things keep getting bigger and bigger. For example, in* The 9-Inch Diet, *you talk about an experience of how you couldn't fit your plates in the cabinets of your cottage, and then it turns out that dinner plates used to be much smaller because portions were smaller. There is a story you often tell from your Dad—that you can always tell who has the most power by the size of their buildings. There's this kind of evolution from the castle to the cathedrals to the state capitals, and now you have these massive office buildings.*

AB: What that illustrates is that the corporations have more power than the nation state. But people don't see that change happened. I think lots of people are nostalgic for the notion that the government is the most powerful force, but I think that's nostalgic thinking. Corporations are more powerful than government—there's no doubt. Last I checked, there was over $20 spent lobbying per vote in the U.S. Your vote doesn't compare to the lobbyist doing an office visit.

Everybody wants a certain kind of world. They've got ideas and concepts that they'd like to see executed through their government.

It's important to realize that you vote with every dollar spent and you vote occasionally in elections. If you want to be an involved citizen that participates in social change, don't give up either of those votes. The voting dollar is a lot more work, unfortunately. It's not every two or four years; it's constant, every day.

DP: That goes back to what you said before around you get what you allow.

AB: Yes. You get what you buy and you get what you fund. You are funding ideals with every purchase, and it's difficult to do the homework to make sure those ideals are aligned. For example, AT&T makes me crazy because they give so much money to people I wouldn't give money to politically. Yet it took me forever to dump my AT&T account, because it's a pain in the neck.

You're funding ideals with every purchase

It's a lot of work for every dollar spent. Who do they give to? What do they support? What kind of policies? What does that mean for gay and lesbian rights? Oh, they're against it. It's too much work for most people vs. going to the polls once every two years and poking a hole. When you see the opportunity to learn something and then make a change, that ability will trend up. If you don't take it, then that ability will trend down.

The bigger issue around everything getting bigger and bigger is that now the global economy isn't expanding. I'm not trying to put the genie back in the bottle with things like artificial intelligence, but no doubt there is a massive disruption coming to jobs. I think we can deal with it economically through things like a basic income. My bigger concern is that *the job* is the number one way we understand ourselves as human beings because we don't really use religion that much anymore.

DP: When I think of all the stuff we spoke about in terms of non-profits vs. for-profit and how to take on the impossible, the work that you did on Truth was incredibly effective at getting people to stop smoking. You used these factoids to educate people about things like ammonia, how light cigarettes aren't better for you, and you had compelling imagery that really had a lasting social impact. Your work with Al Gore on climate change is another great example of meaningful work with social impact.

Hypothetically speaking, if there was an agency for planet Earth and budgets weren't an issue to scale production and creative executions, what type of campaigns and projects would you do? How might the types of solutions that advertising typically provides its clients be utilized to make the work better?

AB: I would focus on online bullying in every form, including stuff that I don't think is considered bullying like trash talk and trolling. People with bully-like mentalities who don't want to hear what other people think use online tools to try to shut down conversations. I think that has enormous costs to society.

I was at a meeting one time at Al Gore's house and Van Jones was there. Afterwards he walked up to my wife and says, "You didn't say anything. What do you think?" My wife's a crazy smart Duke math major, and she said something insightful about climate change. He replied, "The reason why I ask is I learned that the quietest person in the room has the most to offer, so I always seek out what they are thinking."

That's the cost, right now, of having this culture of bullying and outrage that seems to exist with every online discussion. We're wasting the opportunity. We all can talk and not have anyone in between us. We don't have corporations or government or anything that's trying to intervene. We can really talk about stuff that matters to all of us, and that opportunity is wasted right now.

....

ALEX BOGUSKY'S career began over 20 years ago when he joined a 16-person ad agency called Crispin and Porter. Under Alex's direction, Crispin Porter + Bogusky grew to more than 1,000 employees with offices in Miami, Boulder, Los Angeles, London, and Sweden, and with annual billings over $1 billion. During Alex's leadership, CP+B became the world's most awarded advertising agency, and in 2010, Alex received the rare honor of being named "Creative Director of the Decade" by *Adweek* magazine.

Always drawn towards a cause, Alex created groundbreaking initiatives such as the "Truth" campaign, which was named the most successful social advertising campaign in U.S. history. His work with Vice President Al Gore debunked the notion of "Clean Coal." And in 2011, Alex rebranded The Climate Reality Project and launched 24 Hours of Reality—the most highly viewed streaming web program of its time.

Since leaving CP+B in 2010, he and his wife, Ana, keep busy investing and advising in over 40 advertising and tech firms including his investor/advisor role for Lyft, the ride-share platform; investor/advisor in Fearless Unlimited, an agency dedicated to positive social change, where he continues his work on climate change with Al Gore; investor/founder in MadeMovement, an ad agency with a focus on clients that create U.S. jobs; investor/advisor in Humanaut, a marketing agency in Chattanooga; investor/founder at HealthySkoop an organic plant-based nutrition company; founder of Boomtown, a tech-startup accelerator; and general partner in Erli Ventures, an early-stage tech fund.

PART IV

THE ECONOMY OF THE FUTURE

How will crowd-based capitalism reshape the economy?

What are the costs of pursuing unsustainable growth?

How can we leverage technology for abundance?

THE SHARING ECONOMY

Arun Sundararajan

Arun Sundararajan delivers groundbreaking analysis in *Sharing Economy* by revealing the profound transformation and underlying economics of what he calls "crowd-based capitalism." He makes a compelling argument that peer-to-peer commercial exchanges could supplant traditional corporate-centered models, effectively becoming the dominant model in a new global economy.

- How trust enables crowd-based capitalism
- Digital Darwinism and the future of work
- Why the blockchain is like the early Internet
- Decentralization can turn providers into owners

As a Professor at NYU's Stern Business School and Member of the World Economic Forum's Global Futures Council on Technology, Arun backs up his bold analysis with serious economic credentials and extensive research. I love his ability to connect aspirational ideas about sharing and collaboration to a new peer-to-peer economic model that appears superior to industrial capitalism.

DP: The Sharing Economy *begins with a story of you looking through Mary Meeker's slides of annual predictions, which have become infamous for forecasting the future of tech. She foresaw an asset-light generation that wants everything on demand, shifts from access to ownership, flexible work hours, etc.*

You describe this kind of a-ha moment where you realize there was much more going on. In fact, this signified a radical shift where the crowd would eventually replace the corporation as the center of capitalism. Can you tell us about this radical shift towards crowd-based capitalism?

AS: The notion of an asset-light existence focuses on shifts in consumption and the idea that we are going towards a more on-demand, collaborative form of consumption—where you're not buying music; you're streaming it. You're not buying a car, but ordering it on-demand. That gets rid of assets in favor of experiences, or replaces ownership with access. But that is one slice of a broader transition.

The mechanism of producing the world's goods and services at the end of the 20th century was very organization-centric. A lot of large institutions employed people full-time making a set of products and services that were then delivered to consumers. Now we are seeing the creation of systems that draw on resources as needed from a crowd of distributed providers.

Crowd-based capitalism signifies a fundamental shift in how we organize the world's economic activities

For example, there is the Airbnb model where you draw on a crowd of distributed homeowners to provide short-term accommodation; or a crowd of distributed Lyft or BlaBlaCar drivers to get transportation on demand; a crowd of individual investors to assemble small business loans at Funding Circle. More broadly, all of this signifies a fundamental shift in how we organize the world's economic activities. I refer to this as the rise of "crowd-based capitalism."

This is a new system of capitalism that is a hybrid between the 18th-century, Adam Smith peer-to-peer marketplace of the small shop owner, and the 20th-century, industrial scale corporation. Platforms do some of the things that a traditional organization might do—accumulate demand, create uniformity across the customer experience, provide trust through branding—but then they facilitate scale through this kind of crowd-based supply across a wide variety of goods and services.

The economic fundamentals of crowd-based capitalism seem fundamentally superior. A lot of this activity was falling under the umbrella of the sharing economy or peer-to-peer, which seem to connote a philosophy of commerce that is non-capitalistic, whereas what I see emerging is decidedly capitalist. Thus the title of my book is called *The Sharing Economy* but I refer to this broader, fundamental shift as the rise of crowd-based capitalism.

DP: This hybrid new model of crowd-based capitalism that you describe in economic terms parallels the Peers Inc. model of company creation from Robin Chase, where "crowd-based" and "Peers" draw attention to peer-to-peer platforms leveraging the power of the crowd, and "capitalism" and "Inc." underscore the scale and benefits of the modern industrial economy.

One key concept in your approach is the notion of trust. The crowd builds trust through things like relationships and ratings systems, and industrial brands build trust through providing reliability and quality at scale. I read that your interest in trust led you to study the sharing economy. Can you tell us about why trust is so important in crowd capitalism, and explain this revolutionary idea of a new trust infrastructure?

AS: I have been interested in trust for as long as I've been a student of business because I believe the trust systems of a society determine the possibilities for economic exchange. We used to live in tight-knit communities where the only people you could trust were the people who lived in your village and they were your only trading partners. Over time, governments started providing some form of trust by certifying food or creating common currency. Then countries created institutions that granted things like the ability to sign a contract and go to court if it was violated, or the notion of property rights. These trust-based systems expanded trading rights and became drivers of economic growth.

If you think of trust today, you don't write a contract when you buy small things like a cup of coffee. Instead, you trust a brand, which signifies a commitment by a company that "I will consistently deliver a high-quality service and stake my reputation and the future of my business on it." This is why you comfortably drink a Coca-Cola when you're in a country where you don't know the food safety rules, or you ride the roller coasters at Six Flags, but hesitate if the roller coaster was on the side of the highway.

We expanded the economy each time we created these new bases of trust. From the time of eBay, it seemed like we were seeing the genesis of a new system of trust that could expand economic exchange more significantly based on either digitized information about you or digitally represented information about the experiences. The creation of this digital trust infrastructure is fundamental to why we're going to see the shift to crowd-based capitalism.

DP: Throughout history there have been these forms of innovation like currency, government institutions, contracts, and brands that add incremental layers to trust infrastructure. You describe the emergence of how community and peer-to-peer platforms create a new trust infrastructure that has six different parts:

- *Government Certification*
- *Third-Party Certification*
- *Platform (or Brand) Certification*
- *Digital Conduits to Digital Traits*
- *Digitized Social Capital*
- *Digitized Peer-to-Peer Feedback*

Can you help us navigate the shifting landscape of the digital trust infrastructure?

AS: Let me go through each of the items. Government or third-party certification involves having someone other than the contracting parties inject a credible quality signal into the transaction—like the Taxi and Limousine Commission saying that this taxi is safe to ride in or a background check on the driver. Brand certification has to do with a company saying, "You can trust that the quality of this product is good because it bears my name and I have a reputation for high-quality provision."

Institutions and contracts have to do with sets of terms that you know will be adhered to and the possibility of recourse through courts if contracts are violated. They make you more comfortable to buy from suppliers. A license is like a contract. For example, we sign contracts when we buy music from iTunes or books from Amazon. In some sense, that has to do with the continued existence of trust cues in the pre–crowd-based capitalism world or the pre-digital world.

There is still a government-injected form of trust in many of the transactions that we engage. For example, Uber administers background checks and has guidelines that shape what drivers can and can't do. Every smart sharing economy platform invests heavily in their brand because they realize that while the lower digital layer is getting stronger, we are still a society where people trust brands and that's an important gateway to trying a new service.

DP: This relates to the idea that people conducting exchanges on crowd platforms benefit from something similar to a franchising model. For example, when they advertise their apartment on Airbnb, they benefit from the halo effect and trust of Airbnb's brand, community, user experience, and platform the same way that a franchisee benefits when they franchise a business. There is this kind of standardization of quality and experience provided by the crowd-capital brand.

AS: Yes. They benefit from the good experiences that other people had with the same branded experience. The investments made in getting people familiar with the model provide a sense of reassurance that there's a large company backing this transaction. This type of brand consistency and service is similar to the franchising model.

DP: Providers also organize communities and standardize their own best practices. For example, Douglas Atkin, the Global Head of Community for Airbnb, describes how their communities of hosts self-organize around the world. In the digital trust infrastructure, there is this layer of peer-to-peer community that basically standardizes delivery of products and services in addition to the standardization and trust from the brand itself—it's all mutually reinforcing, going back to what you said earlier in terms of a hybrid between the Adam Smith, peer-to-peer marketplace vs. industrial capitalism.

AS: Yes, I think those kinds of decentralized, guild-like ways of setting standards are going to be an important part of the future regulatory infrastructure. We have some self-organizing, self-regulating components in other industries that have historically been peer-to-peer—like the lawyers have the Bar Association, the doctors have the American Medical Association, and real estate agents have the National Realtors Association. Governments dictate some of the safety of these businesses, but there is a strong community-based, self-regulatory component to setting and enforcing standards of provision. It seems like a natural evolution for the sharing economy as well.

In terms of the digital trust infrastructure, there is basic proof of who you are with respect to providing mobile number, government-issued ID, membership in different organizations, etc. I can hold my passport up to a webcam to identify who I am. We can also add individual interests ("I like jazz music") that could cause someone else on the platform to trust you more depending on shared characteristics.

Looking into the future of where this is all heading, we'll have friends, family, people who know us, and business acquaintances whose presence signals a particular set of characteristics that you are an effective or trusted member of the community. We already digitize and make available our real-world social networks on platforms like LinkedIn and Facebook, so our real world social capital will become part of trusted digital interactions. For example, I look at someone on Airbnb or BlaBlaCar and discover that we have LinkedIn or Facebook friends in common. That plays a role in facilitating trust.

Then there is the digitized peer feedback, which is based on the rating and reputation systems invented by eBay many years ago that were the basis for their rapid growth. "abc123 has had 73 successful transactions and a 4.9 star rating." By itself, it is unlikely to be a sufficient basis of trust for a more high-stakes interaction, like you can sleep in my spare bedroom or get into my car and I'll drive you to another city, but it was sufficient for receiving a package from someone in the mail.

Our real world social capital will become part of trusted digital interactions

High-stakes peer-to-peer interactions require the creation of the rest of the digital trust infrastructure, but they also require some reliance on existing trust cues, like brand. For example, I'm doing a study on BlaBlaCar. Early on, users place a high weight on the brand as a way of facilitating trust. As they gain experience, they rely more on different parts of a digital profile. Some focus on peer feedback, others care more about "Oh, this person is an Auto Club member" depending on what syncs best with their notion of trust. Brand is an important dimension of onboarding people, after which the digital profiles kick in as an important basis for continued interaction.

DP: I'd like to shift directions to talk about how to best scale crowd-based capitalism and peer-to-peer platforms. Your book opens discussing some tensions at OuiShare Fest among people that think about the sharing economy primarily in the context of a collaborative society. I've been to OuiShare Fest before and I also interviewed Antonin Léonard, the co-founder of OuiShare for the book.

On the one hand, there are many passionate supporters of collaborative and sharing platforms motivated by altruistic goals to change the world. Investors can help these initiatives to scale faster, and there is certainly a movement among many investors to only back companies that want to make a positive impact on the planet. Yet on the other hand, the massive infusion of venture capital into peer-to-peer companies creates a situation where investors expect massive returns on investment that could cannibalize the egalitarian nature of these community-based platforms.

This scenario reminds me of Timothy Wu's book, The Master Switch, *where he outlines this pattern of super passionate innovators and early adopters advocating the world-changing potential of emerging technologies like the telephone, the radio, TV, and cable, i.e., humanity will be better connected than ever before, creating a new era of human civilization. Then business people come along that transform these innovations intended to benefit humanity into monopolistic empires through ruthless competition. The 'master switch' led to scale, but it comes at the expense of losing the vision to help humanity.*

One could argue that we see the same type of pattern around dominant tech companies like Facebook, Google, and Apple, and again with the sharing economy. How do you think peer-to-peer companies and crowd-based capitalism will manage this balance between serving their communities vs. investors? Will we see a wave of disruption—a kind of sharing economy 2.0? How do you see the future?

AS: I think it's a very good parallel. Looking at the sharing economy in isolation, you do see the frustration. But looking at it from the historical perspective, the frustration among the idealists of the large capitalist machine taking over is a familiar story.

In terms of the future, I expect that as we continue to exist in the world of intermediated peer-to-peer (the world of Kickstarter, Etsy, Airbnb, and Uber), we are bound to see more variety. Some platforms will be more investor-focused and others will focus on users. This may have to do with the founding philosophy or the nature of the industry.

For example, look at Airbnb and Uber. Airbnb founders have backgrounds in design and began the company as hosts designing quality hospitality experiences. The genesis of the company was through the eyes of a host. Uber founders had a background in tech and finance. They started the company wanting a better experience for the customer instead of the provider. Nobody complains about Uber's customer experience, but Uber drivers can be less than fond of the platform. On the other hand, Airbnb's hosts seem to love the platform, even self-organizing communities and sharing with each other.

The emphasis on community ideals vs. investor objectives is shaped not just by the nature of investment or capitalism itself, but different companies have different personalities and each strikes a different balance.

DP: This is an excellent point regarding the founding philosophy and nature of the industry, one that is also raised by Douglas Rushkoff and Michael Bronner. It also points to a broader shift in values, where the short-term wins accelerated by massive investment of venture capital may not be sustainable in the long-term with next-generation platforms designed specifically to share ownership with providers.

This leads to questions about platform design and decentralization. There is a lot of hype and speculation around the possibility of decentralized platforms built on the blockchain. We are still in the early stages of this technology and it's unclear where the real opportunities lie. What are your thoughts on the blockchain?

AS: The excitement around the blockchain today reminds me of the excitement around the Internet when I was a grad student. This is a liberating technology that will bring down governments and eliminate the need for intermediaries. It will free the world. It will encapsulate everything in code, and existing institutions will crumble. This sounds a lot like the Internet in 1995 through the eyes of the idealist.

It's a pretty revolutionary technology, but I'm not a technological determinist in the sense that I don't believe the properties of the technology itself shape its role in society. We should think about technology in conjunction with existing social structures, what human beings are like and how they form collectives, how some desire power more than others—all of these will play a factor in shaping what impact blockchain technologies eventually have on our institutions.

DP: It's great to remind us of how important the human element is in the equation, which brings us back to the importance of collaboration and the crowd. If society shapes the role technology plays, what factors should we take into consideration to evaluate decentralized platforms and the blockchain that may create a kind of sharing economy 2.0?

AS: In the near term—or the next 10 to 15 years—there are a number of different reasons why we are likely to see the continued existence of intermediaries in the sharing economy. Any time there is a decentralized system that can potentially connect individuals with other individuals or eliminate the need for an intermediary, there are capabilities that the users look for that are sometimes not adequately provided by the decentralized system. I'll give you three examples.

We are likely to see the continued existence of intermediaries in the sharing economy

First, think of the early days of the Internet. Conceptually, the web is the ultimate decentralized publishing platform. Anybody can publish something that anybody else can read, so it gives everybody an audience of the world. Very soon you realized that it is hard to find what you were looking for and to credibly decide whether the stuff that you found was any good. Then came search engines.

Google re-aggregated this distributed value creation. They became an intermediary in a world where it seemed unlikely that we would need one because anybody could publish. They don't extract all of the value, but they capture a significant fraction of it because decentralized systems require some intermediated search and discovery.

By analogy, if we start to see blockchain-based peer-to-peer systems for a variety of goods and services—journalists publishing and trying to capture a greater fraction of their revenue, decentralized peer-to-peer versions of ridesharing, peer-to-peer accommodations—we're still going to see an intermediary re-inject itself and say, "Well, I can do better search and discovery than the system can."

Second, intermediation has a similar advantage with anything that requires logistics. Bitcoin is convenient because it's money. Payment is just the exchange of a relatively small amount of information to clear a transaction and then you're done. Once you start to get into the trading of physical products or the delivery or real-world services, there are real world logistical challenges to deal with.

For example, Alibaba is the dominant retail platform in China. It's four times the size of eBay and much bigger than Amazon. It's fundamentally peer-to-peer, but they rely on a set of third-party, closely tied logistics providers who handle the delivery. That has not been decentralized. Similarly in the dispatch of cars, even if we see the genesis of decentralized, blockchain-based ridesharing systems, I think they will be re-intermediated by someone who says, "I can help you do the logistic setup."

Finally, there's the issue of trust. I realize that the blockchain is positioned as the ultimate basis of trust because you can track prior transactions. But as we discussed, creating trust in a peer-to-peer environment is a complex and multi-faceted thing that involves a wide variety of cues, some of which lends themselves well to a centralized intermediary. I don't think that as a society, we are ready yet for population-scale, decentralized exchange without some sort of injection of trust from either a third party provider, like Uber, or some sort of platform, like Airbnb.

For a peer-to-peer alternative based on the blockchain to emerge for properties like Airbnb, I think it will be a while before they can overcome the fundamental advantage of an established platform that has built trust and has good trust mechanisms simply because they are cutting out the fee that the intermediary collects. Airbnb's commission is a thin margin, anywhere from 3% to 15%. If I can save like $10 on a $100 transaction but I have to take a bunch of risks, I may not be that inclined to gravitate toward the new peer-to-peer solution. Uber's margins approach 40% for SUVs in New York and 35% for car service, but those margins will come down through competition.

The real potential is in communities where the providers currently get an extremely small fraction of the revenue that their creation generates. For example, in the music business or journalism, you have intermediated structures where the creator gets a very tiny fraction of the value that they create. They are also information-based products and services, so they don't have the logistical challenges of delivery, and the trust infrastructure is a lot easier. Blockchain-based intermediation will probably gain traction there or in financial services before it makes inroads into the real world goods and services that are currently the focus of the sharing economy.

The real potential of the blockchain is in communities where the providers currently get an extremely small fraction of the revenue that their creation generates

DP: *The subtitle of your book includes the phrase "the end of employment." I also interviewed Martin Ford, author of* Rise of the Robots: Technology and The Threat of a Jobless Future. *The consensus from most of the book contributors is that employment in traditional career paths is basically over.*

Can you tell us about the challenges and opportunities you see with the end of employment?

AS: The end of employment sounds dire, but it's actually a sort of optimistic prognosis for the following reason. The role of the individual in the traditional employment relationship is one of labor provider for a wage. The labor might be intellectual labor or professional, but the fundamental construct is that you give me your time and expertise and in exchange I will pay you a fixed wage. The institution remains the owner of the system of production.

What we're moving towards is a more decentralized system in which the role of the individual shifts from being a provider of labor to an owner of sorts. An Uber driver is less of an owner than an Etsy seller, but fundamentally at the heart of it is the shift away from provider of labor to owner of something. An optimistic view is that the end of employment corrects imbalances inherent in the employment relationship.

The employment relationship was popularized 100 years ago and went through a process of creating perfections, wrapping a safety net around it, and facilitating stability and quality of life that individuals aspire to have. Before that for thousands of years, most work was basically peer-to-peer in small local communities. Fear about the future of work is governed in part by the fact that we are comparing the employment relationship to a new freelance, micro-entrepreneurship model that hasn't gone through a similar refinement process, though I think that will happen in the next 10 to 20 years.

At the heart...is the shift away from provider of labor to owner of something

DP: In an interview for the a16z Podcast, you also spoke about the emergence of a hybrid-firm marketplace model, which I thought was really interesting. Contently is an excellent example of a company that acts like a firm providing services to big brands, and is also a talent marketplace for over 50,000 journalists. I spoke with Shane Snow, a co-founder of Contently, for the book.

One of the things that I found interesting about your take on the future of work was this idea of Data Darwinism. You speculate that a day may come where the same types of ratings and review systems applied to products on platforms like Amazon could be applied to people in the on-demand workforce.

There is the potential for new industries to emerge managing workers' data and ratings across multiple on-demand platforms. You also talk about how the autonomous world of drones and self-driving cars could enable new forms of employment or direct peer-to-peer exchanges.

Can you tell us about the hybrid firm-marketplace model and the idea of Data Darwinism?

AS: Yes, and in some ways, these will be the new institutions. There are consultants or individuals who provide labor and run their own businesses. What we'll see emerge is some combination that looks like freelance work and also a bit like entrepreneurship. They will play an important role being actors that help us construct this new social safety net.

As we transition to this world where more and more people are individual providers rather than full-time employees, a lot of what will dictate the opportunities that an individual provider has will be their reputation—their portfolio of prior interactions, much like Airbnb host prospects are determined by the feedback from other people on the Airbnb platform. If you're a consultant, then ratings from prior clients will dictate your future opportunities.

The idea of Data Darwinism is originally from Om Malik. He made an observation during the first Uber driver protest when a bunch of drivers were shut off from the platform because their ratings went below a certain level. He imagined what happens in a future when there is a new form of survival of the fittest where if I get off to a bad start, then I can't compete with the other people who started off with stronger reputations. You get this new kind of Darwinistic selection in the workforce.

Darwinism is not necessarily a bad thing. Certainly we have seen a lot of good outcomes through the last few billion years of dominant selection. What makes me cautious is that the selection is going to be based on assessments of individuals by other individuals. They are subject to the same kinds of biases we have when we're making subjective assessments. People may systematically and unconsciously evaluate others of a particular ethnicity, race, gender, religion, etc. ,higher than others.

We know that bias exists in society, and these rating systems may not be sophisticated enough in their simplest forms to capture and account for this kind of bias. You might see the propagation of disadvantages that have existed because of bias in traditional society into this world of ratings. It's important for anybody running a system like this, whether it's across multiple platforms or within a certain platform, to detect and correct for any sort of bias.

ARUN SUNDARARAJAN is Professor and the Robert L. and Dale Atkins Rosen Faculty Fellow at New York University's (NYU) Stern School of Business, and an affiliated faculty member at NYU's Center for Data Science and Center for Urban Science and Progress. His new book *The Sharing Economy* was published by the MIT Press in June 2016. He has published over 50 scientific papers in

peer-reviewed academic journals and conferences, and over 30 op-eds in leading outlets globally. His scholarship has been recognized by six Best Paper
awards and two Google Faculty awards. He is a member of the World Economic
Forum's Global Agenda Council on Technology, Values and Policy, and advisor
to numerous other policy bodies, venture capital firms and cities. He has provided expert input about the digital economy as part of Congressional testimony,
and to a range of government agencies. He is a widely sought-after commenta-
tor by top media platforms.

THE GROWTH TRAP

Douglas Rushkoff

Douglas Rushkoff makes a compelling case in his latest book *Throwing Rocks at the Google Bus* that we are caught in a growth trap—pursuing an unsustainable model of unlimited growth, locking up valuable capital in stock prices and seeking 100X returns on investments often at the expense of cannibalizing the health of good companies.

- Reprograming the system from the inside
- Mass conformity in a sea of unlimited choices
- Why CEOs should optimize for revenue
- Founding companies to keep instead of sell

Douglas presents concepts like corporatism, money, and the economy in a way that makes them easily accessible and fun to talk about. This is probably why his numerous bestselling books appeal to activists and Fortune 100 CEOs alike. His common-sense approach will draw you in and leave you wanting to read more. After his latest book, *Present Shock* and *Program or Be Programmed* are my personal favorites.

DP: There is a recurring theme throughout your writings to question or reprogram the basic operating systems of society, like corporatism and money, to benefit people instead of extract value. I love this one quote from your latest book:

We need to make a choice. We can continue to run this growth-driven, extractive, self-defeating program until one corporation is left standing and the impoverished revolt. Or we can seize the opportunity to reprogram our economy— and our businesses—from the inside.

For people unfamiliar with your work, can you explain what you mean by an operating system and reprogramming from the inside?

DR: The operating system is what makes your computer go. There's the Windows operating system and Macintosh operating system, and all the software that we use runs on top of that. Sometimes it's easy to forget that you are building software for a particular operating system—that the operating system has biases, ways of working, underlying assumptions about the way people interact with machines, what the role is of computers in peoples' lives, and all that.

If you were to wake up in a world where there was only the Macintosh, you wouldn't even know that there is such a thing as an operating system. You would look at your computer and say, "Well, that's just a computer." You wouldn't understand that there's a choice of what underlying system to use for building your software. I think it's a fitting analogy for business and the economy.

For example, most entrepreneurs and investors look at the VC startup mechanism as the only operating system available. They don't even see an operating system—it's just the laws of business, a fixed reality. The natural condition of things is that you take money from these angel guys, you do a Series A, then a Series B, and you go for an IPO and try to return 100x or 1000x to your investors. As if it's like, when a cell undergoes mitosis, it has four stages, or a woman gets pregnant and then there are these trimesters and the baby comes out. As if that's just the way things are. And it's not.

The VC startup mechanism is a very particular way of doing business that was invented at a particular moment in history meant to leverage digital innovation towards very specific ends. What it does is trap potentially innovative, sustainable, breathtaking businesses in a self-destructive pattern. The obvious example is a company like Twitter, which can make $500 million a quarter or $2 billion a year on a 140-character messaging app, and it's considered an abject failure by Wall Street because the operating system on which they're running is a growth-based operating system.

DP: This gets into what you refer to as a growth trap: the idea that the rules of the operating system of capitalism and venture capital trap companies into raising so much money to maximize growth that they are unable to pursue sustainable business models. They need to ramp up fast to gain a monopoly and have a big home run for investors at 100x or 1000x returns, becoming trapped by a relentless focus on growth.

DR: Yes. The trap is when a company has to grow and the capital gain of the shares is more important than the revenue opportunity and long-term prosperity of the underlying business. It's even screwed up the stock market which is growth based. Consider the derivatives market—which is able to show growth in advance, to basically time-shift your growth—is so much more important in this schema that the derivatives exchange purchased the New York Stock Exchange. The stock exchange was eaten by its own abstraction. That's how important growth is. Growth trumps the health of the underlying business.

What happens if a company realizes it's reached a sustainable but ultimately limited level of revenue?

The growth trap is part of a financial program based on selling the company again and again at larger valuations. This is the same as during the housing boom when people would buy houses and finance them by refinancing repeatedly at greater and greater valuations. What happens if a company realizes it's reached a sustainable but ultimately limited level of revenue? What if a company can only make $2 billion a year?

Well, for Joe's Pizzeria, that's not a problem because Joe doesn't have investors to pay back. He's not running that same operating system. He's running on the family business operating system rather than the financial services operating system. He doesn't have to think about that. Joe can be great on making a billion dollars a year—that's so much money—but for Twitter that is a problem.

Even though Twitter can easily pay their employees good salaries and their original investors could make great dividends quarter after quarter and year after year, that's not the plan. That's not the way that this operating system works. Sustainable, revenue-based digital businesses are incompatible with the underlying growth-based operating system that, as I understand it, was put in place in about the 12th century to support the colonial expansion of Western European nations. To get out of the growth trap, we need an operating system to support the sustainable equilibrium of a digital economy.

DP: Let's dig into the origins of this stuff in the 12th century. I have a background as a historian of religion and culture, and one of the things that I love about your latest book, Throwing Rocks at the Google Bus, *is how you frame the rise of industrialism in the context of aristocrats asserting their dominance over peer-to-peer trading in the Middle Ages.*

You paint a picture of how local merchants used to gather in a bazaar that looked something like Burning Man. There were guilds that regulated trade and allowed people to make collective decisions. The prosperity from these groups basically threatened aristocratic power, so they intervened and set up a new system of colonial control. Can you tell us about how things used to be and the decisions made centuries ago that led to the creation of industrialism/modern-day capitalism?

DR: First, I should say that things weren't great for long periods of time. You have to sift through history to find these bright spots of minimal exploitation. People lived for many centuries under feudalism, which really—there's no nice way to say it—just sucked. They worked the lands for the vassals of the lords and it was horrible. Then men were sent out on the Crusades, which also weren't fun. Everybody died and killed and it was an awful thing.

The Crusaders encountered different cultures around Europe and the Ottoman Empire, and when the Crusades finally ended, two big things happened: First, they had opened trade routes through their military conquests; and second, these soldiers came back to their towns throughout Europe having seen all of these other ways of doing things around the world. Whether it was new ways of milling one's grain, structuring a bridge in a waterway, or economic innovations like the bazaar.

The bazaar was an Arabic marketplace. European towns started to imitate the bazaar, which is where people would bring stuff that they made, fabricated, farmed, or killed. One or two days per week, everyone went to a central location and exchanged goods and value added services. Barter was an inefficient way for people to trade with all this stuff, so they developed and borrowed a number of financial innovations that they encountered in the Ottoman Empire.

There were things like market money and different kinds of grain-based currencies, commodities-based currencies, and even currencies based on almost nothing (like poker chips that were issued in the morning just to promote, prime, and pump for trade). The economy was optimized for the exchange of goods and services. It became so extremely efficient that people began to work less. Women got very healthy because they were eating so well. Women were taller in medieval England than at any time until the 1980s, that's how healthy they got.

There was still awful stuff—rats and rapes, for example—but you saw huge economic growth as the former peasants became the middle class. A new rising merchant class emerged called the bourgeoisie, and the aristocracy didn't like this because they enjoyed less power. They couldn't absolutely control the entire populations of their countries with a rising and prosperous merchant class.

The aristocracy implemented a few innovations to stamp out what today we would refer to as a peer-to-peer economy. One of them was central currency, which made all of the local currencies, grain-based currencies, and market moneys illegal. The other was the chartered monopoly, or what we now think of as corporations. You weren't allowed to do business in a particular industry without the king's official sanction. He only gave that to one or two companies in each sector, so they enjoyed monopolies by law.

Monopolies in industries killed competition and invented the notion of working for time wages, or what we call employment. Instead of being a craftsperson, shoemaker, or ironsmith, now you worked in a factory for someone else. You became an employee of His Majesty's officially sanctioned smithy. That ended the peer-to-peer economy pretty quickly and Europe went into poverty again. That's when we got the Black Plague and what we now like to call the Dark Ages.

DP: When the Internet came along, the initial promise was similar to the bazaar. I interviewed Chris Anderson for my last book, who wrote The Long Tail, *which laid out the promise of how millions of people could find niche markets that service their unique needs, and millions of obscure artists and musicians could make livings from their crafts, selling goods to audiences, and so on. They might not get rich, but they could make a healthy living selling to customers thanks to the long tail.*

One of the things that I was surprised to learn reading your book is that the long tail hasn't enabled the type of prosperity among the outliers that we all hoped it would. In fact, the data shows a few blockbuster hits make up a greater percentage of music sold than before the days of the Internet. For example, in the past 80% of sales came from 20% of products, and today the bottom 94% sell fewer than 100 copies each. Why in a sea of seemingly infinite choices do people overwhelmingly choose the same thing? How does so much freedom lead to a shocking amount of conformity?

DR: The main reason is that Internet platforms have self-reinforcing loops, so if something makes it to a top ten list on iTunes or top 100 list, those are the only things that many users see. The algorithms are not configured to maximize the distribution of the long tail amongst as many players as possible. They're optimized to create self-reinforcing loops that maximize the numbers of sales.

For example, once Taylor Swift shows up on a list, it's going to self-perpetuate. It's the difference between 100,000 record stores in America each deciding what record they're going to play during the day to influence shoppers and the same record being emphasized to the entirety of the world on iTunes. The more centralized and closer to a monopoly, the more influence a platform has on selections.

The more centralized and closer to a monopoly, the more influence a platform has on selections

People tend to flock. They aggregate around things. If you're out on the Internet, then you're basically alone. The only way to forge connections with other people is through affinity, which is not the same thing as community. Affinity is just "Do we like the same stuff?" Affinities create self-reinforcing loops. "Oh, she likes Taylor Swift. I like Taylor Swift." Then, you're going to buy Taylor Swift.

DP: Lots of people champion social media for basically connecting humanity greater than ever before. We spend countless hours on platforms like Facebook, Twitter, Snapchat, and Instagram, sharing and interacting with friends and family. But according to your model, we are all actually the products of social media companies. Every exchange and interaction becomes a data point that can be packaged and sold. Likes and engagement metrics drive the valuations of social media companies into the billions, even if they don't generate any revenue. Can you explain the idea that we are all products?

DR: The easiest way to understand it is to ask: Who is paying the social media company? How do they make their money? Figure out who is the customer of a particular platform. Most people think, "Oh, this social media platform exists to help me make friends and maintain good friendships." But go into the boardroom of Facebook, Twitter, Instagram, SnapChat, etc., and they are not asking, "How are we going to help little Johnny make more enduring friendships?"

The real business model of social media companies is to get money from marketing companies and market researchers for the data that they can derive from people's interactions. That means the question they ask in the boardroom is: "How are we going to optimize Johnny's social graph?" The activity that Johnny does is actually the product. That's what they are selling.

Instead of needing a market researcher to stick a camera into Johnny's closet, which is what they used to do to see the selection criteria for his clothes each day, now Johnny voluntarily creates a consumer profile. This is how a young person represents him or herself. They work on their profile meticulously, because the movies they like, the books they read, the music they listen to, etc. is how they express their identity to everyone. This is why I say if you're not the customer, you're the product.

It's not so bad if you understand that is the deal, but what most social media users don't realize is that the entire platform is optimized for that. It's not optimized to help them do anything. It's optimized to either market to them or to get them to reveal more and more useful data about themselves. Most people don't understand that things they see on Facebook have been algorithmically prepared to maximize certain perceptions and actions.

Most people don't understand that things they see on Facebook have been algorithmically prepared to maximize certain perceptions and actions

Their feed is not just a representation of what they ask to see. It is a combination of people paying to get their attention and algorithms trying to influence their behaviors. Your Google search results are different from my Google search results. That's because neither you nor I are paying Google for that. Who is paying Google for that? The people who want to influence our search results.

DP: There is a similar challenge within the peer-to-peer economy. On the one hand, you have companies like Uber who view peer-to-peer communities as basically means to an end. They exploit inefficiencies in the transportation ecosystem with temporary contractors that will likely all be replaced one day by autonomous cars. On the other hand, you cite examples like Meetup, Pando, and Kickstarter where the startup founders made choices early on to build sustainable companies that align with their core values. How do you build a company from the ground up so it doesn't get caught in the growth trap?

DR: The easiest way is not to take too much money, particularly from investors who don't believe in the core mission of the company. Now, I understand when you're starting a business that you want to have a lot of cash around. It feels more secure. The problem is when you have a lot of cash around, it's tempting to see investors as the customers rather than to focus on your real customers.

> *When you have a lot of cash around, it's tempting to see investors as the customers rather than to focus on your real customers*

The key is to find a way for the revenue that you make to support the operations of the company. If I was making pizza, then I would want to figure out the costs of materials, labor, rent, etc., and how much I can charge to make a profit. Then you try to feed profits back into the business and grow naturally. If the pizzeria had taken a million dollars upfront from the mafia, it needs to pay the mafia a million dollars. That will not be through slow revenue over time. That's going to be from exploiting something or letting someone use your restaurant for a poker game at night—the pizza be damned.

DP: On the other side, most companies have shareholder agreements that basically mandate the leadership team to yield high returns on investments, effectively legislating the growth trap. This forces companies to extract as much value as possible from customers and then lock it into their stock prices. Most shareholders actually prefer higher stock prices than dividends, so that money isn't circulated back into the economy.

In your book, you have a term called the "steady-state CEO" that could be an advocate for change within the narrow confines of most companies. Can you explain the idea of the steady-state CEO, and for people working within companies, what type of options do you see available to them to implement some of the changes that you describe?

DR: The idea would be to not require a company to grow in order to be considered successful. For example, what happens when you are a multi-billion dollar company in the Fortune 100 and your investors are demanding gross on the bottom line? In many cases, the CEO sells off the company's productive assets in order to show short-term growth.

The idea would be to not require a company to grow in order to be considered successful

The reason they sell them is: (1) Productive assets create revenue and revenue is a problem because they have to pay tax on it, and (2) You can make a lot of money because the parts of the business generating revenue are the ones that people want to buy. These are reported as capital gains, so they are not taxed the same way as revenue

As companies keep doing that, they end up sitting on a mountain of cash gained through selling aspects of the businesses that are productive. This inflates the value of the stock, making shareholders happy. However, it becomes self-cannibalization. You bring home meat by cutting off your own arm. It also leaves companies incapable of showing new profit.

This is why Deloitte found that over the last 75 years, corporate profit over size has been going steadily down. It's not just a digital thing but a business problem. Corporations got very good at accumulating cash but have gotten increasingly worse at deploying the cash that they collect. Ultimately, chasing growth in that way kills your company.

What CEOs need to do instead is optimize their companies for revenue. That means accepting it's OK to be a certain size. It's OK to be just a $2 trillion company or whatever its valuation is. Then run the engine rather than try to figure out new ways to expand or new places to extract money. Even if you end up operating at the same level for decades, you can still return tremendous money to shareholders.

The problem is that shareholders don't want dividends because the tax is biased against dividends. The tax system was lobbied for and put in place by growth-based CEOs and lobbyists of growth-based companies. They kind of jerry-rigged the tax system to favor the way in which they made money over the last century or so, but it really works against the way they should be making money now.

What they need to do is lobby for a flip of the tax system so that capital gains are taxed high and dividends are taxed low. They need to turn their shares into something more like preferred shares that remain stable over long periods of time but can deliver 7/8/9% dividend returns. That's not your growth-based, screaming 100-1000x return on investment, but it is a healthy sustainable return. Not every company can be perpetually in its growth phase.

DP: Given all of your writing about operating systems, corporatism and money, the decentralization of power and value exchanges, the importance of flow over growth, etc., it shouldn't be surprising that you are optimistic about the potential of blockchain. There's so much wide speculation around what the blockchain could enable, from new types of currencies and decentralized marketplaces, to new types of companies with collective owners that are completely transparent and open. The biggest enthusiasts speak like it's a panacea for all of our problems, and then that ends up attracting investors that may perpetuate the growth trap. What potential do you see in the blockchain?

DR: People have to realize what the blockchain is for and what it's not. The blockchain is not a panacea. It is not something to use in a 500-person company that exists in a particular town. Nor is it something that you need in your community in order to have a global currency or to figure out your local tax rate. Bitcoin is certainly not a panacea. Bitcoin is, as I see it, a digital currency for gold bugs. It recapitulates all of the problems of a scarcity-based central currency, only it also uses a ton of energy and it favors legacy players, people who have been mining it with computers since the beginning.

Blockchain is interesting for global-sized collaborative enterprises and giant projects where people don't trust each other. For example, if you are doing a giant Wikipedia-sized project where people are going to get paid based on how many articles they've written, or if you have a global Uber of cab drivers that's owned by the drivers and they want a way to verify that people drive commensurate with their share of ownership. Then you can use a blockchain as a way of doing a project in a decentralized way. Or when the automated clearinghouse takes too much money for its transaction services, then people might turn to a blockchain because they no longer trust the authority to authenticate what's going on between them. But to be clear, there is a premise of distrust to begin with here.

I personally don't have a problem with centralized leadership as long as it is honest and transparent, rather than exploitative and opaque. The mere existence of blockchain technologies and the possibility of that alternative may keep central authorities in a certain kind of check. The fact that we don't need them in order to negotiate with each other or verify may be enough of a deterrent to dishonest practices among central authorities.

The mere existence of blockchain technologies and the possibility of that alternative may keep central authorities in a certain kind of check

I feel bad for the DAO Ethereum project that got hacked because they were idealistic people looking for a different future. But once you see the Winklevoss brothers investing in something, you should realize there's a problem because as you said, it means traditional investors are trying to game the system. They turned Bitcoin into a traditional investment rather than what we were thinking it would be, which is a decentralized verification system. The challenge with any project built on top of the blockchain is to keep that from happening again and again.

DP: The future can appear to be a little daunting when we think of the potential for joblessness from automation and all of the challenges that we face due to the monolithic and singular focus on growth. Yet in your writing and speaking, you remain optimistic—"We're all in this together" and "We don't need to be trapped in a future that nobody wants." How do we stay hopeful and persevere to get the well-being and security that we all deserve?

DR: I think we need to accept that there is no free lunch and the easiest way to make money is to earn it. Most people look at their jobs and investments as if the object of the game is to make enough money so that you don't have to work anymore. What I'm trying to convince especially developers and entrepreneurs is that it's OK and noble to work for a living.

Maybe you should found a company because you want to keep the company, instead of founding it for the purpose of selling. I know that sounds radical or socialist, but it shouldn't be. You can build an application that you love and want to serve people. This is how you make companies that end up serving markets rather than killing them. Amazon killed the book industry because it wanted a monopoly. Uber is killing the driving business because it wants a monopoly to leverage into something else.

Maybe you should found a company because you want to keep the company

This will feel better. You're going to wake up happier in the morning if your business promotes and serves a marketplace rather than destroying it. Yes, you'll have to work. You'll have to serve. You'll have to keep doing stuff. But that's fine. You're still alive. If you enjoy your work and you're not seeing it as a mean, exploitative thing, then it won't feel like such a cop-out serving a market. You won't have a sense of failure about creating something that you need to keep working on because you believe it's a good and noble thing to do.

....

DOUGLAS RUSHKOFF is a writer, documentarian, and lecturer whose work focuses on human autonomy in a digital age. He is the author of 15 bestselling books on media, technology, and society, including *Program or Be Programmed*, *Present Shock*, and *Throwing Rocks at the Google Bus*. He has made such award-winning PBS Frontline documentaries as *Generation Like*, *Merchants of Cool*, and *The Persuaders*, and is the author of graphic novels including *Testament* and *Aleister & Adolf*.

Rushkoff is the recipient of the Marshall McLuhan Award for his book *Coercion*, The Jacques Ellul Award for his documentary *The Merchants of Cool*, and the Neil Postman Award for Career Achievement in Public Intellectual Activity. Named one of the world's ten most influential intellectuals by MIT, he is responsible for originating such concepts as "viral media," "social currency," and "digital natives." Today, Dr. Rushkoff serves as Professor of Media Theory and Digital Economics at CUNY/Queens, where he recently founded the Laboratory for Digital Humanism and hosts its TeamHuman podcast.

THE END OF EMPLOYMENT

Martin Ford

Artificial intelligence and robots will initiate a new era of automation that impacts every industry and redefines the future of work. Martin Ford anticipated this trend in his first book *The Lights in the Tunnel* (2009), and his bestselling *The Rise of the Robots* (2015) was McKinsey and Financial Times Business Book of the Year. We explore the tsunami-sized wave of disruption coming to employment:

- Why AI in the future will be analogous to electricity
- Manual labor can be harder to replace than office jobs
- Specialization might make your job easier to automate
- The economic scenarios we might face in 10-20 years

Martin Ford loves technology and doesn't want to be the doomsday guy. He is passionate about starting a conversation on the future of work because the problems society faces will require unprecedented international cooperation and political solutions. Nobody will be unaffected, but together we can co-create the future we want and deserve. Please read and then join the conversation!

DP: The title of your recent book The Rise of the Robots *conjures up images from science fiction with androids taking over and challenging humans for control of the planet. It becomes quickly clear in reading that you refer to a much broader, different type of disruption happening across all industries. What does the rise of the robots look like?*

MF: The word "robot" is used very broadly. Really, what I am talking about is artificial intelligence and automation. Often that is going to be just software. One of the biggest issues I point out in the book is that a lot of skilled jobs held by people with college-degrees sitting in front of computers doing some type of knowledge work are highly susceptible. That obviously has nothing to do with physical robots.

DP: In the late-19th century, nearly half of U.S. workers were on farms, then by the end of 2000 those numbers were less than 2%. The economy adjusted with mass migration of workers to major cities, and widespread access to education led to the rise of a flourishing and prosperous middle class.

There was what you describe as a nearly perfect correlation between increasing productivity and rising incomes. A lot of economists look to the past and suggest that new jobs will be created by new industries and fears of automation dating back to the Luddites smashing textile looms are unwarranted. What is different now?

MF: There will be new industries in the future, but they are not going to be labor intensive. They won't hire many people. Look at Google, Facebook, all of these industries in the last decade or two, none of them hire huge numbers of people proportionate to the massive impact that they have on society. There is nothing out there like the automotive industry, for example, that created an enormous number of jobs, both directly, like people working in factories, and indirectly, in terms of people driving cars. The industries that we are generating now don't look like that. This is what I mean by basic labor intensity.

Next, there is the nature of the work. As you said, most people worked on farms. They transitioned into factories during the industrial era, and then from factories into the service sector. The key thing to understand in each of those cases is that the nature of work was basically routine in nature. People did routine work on farms, then they moved into factories and did routine assembly line work, and they continued to do routine work in the service sector. All routine work—anything that is on some level routine, repetitive, predictable—is susceptible to automation, machine learning, and other stuff.

All routine work—anything that is on some level routine, repetitive, predictable—is susceptible to automation

This is tremendously broad-based and applies to about any kind of job that is on any level routine and repetitive. What that means is in order for workers to still have a job in the future, they have to make a different transition than workers in the past moving from one sector to another, from agriculture to manufacturing. In the future, if you want to have a job, you will have to make a much more difficult transition into something genuinely not routine that may be creative or in some way protected from what machines are capable of doing, at least for the time being.

There is a real question as to will there be enough jobs of that type, because historically we haven't had everybody doing this blue-sky thinking kind of stuff. Most people do routine jobs. The next question is can most people make that difficult transition? I think a lot of people won't be able. Most are probably best equipped to do relatively routine work. Those are challenges.

DP: There is this massive, broad shift occurring across all industries due to automation and AI. Virtually nothing will remain untouched in a sweeping wave of disruption and integration of this technology into the fabric of everything that we do. This reminds me of a section in the book where you refer to artificial intelligence like a utility, similar to electricity or water in the future. What do you mean by that?

MF: AI is extraordinarily broad based and it's going to be applied everywhere. The tools and building blocks for AI and machine learning are going to get better and easier to use. What that means is as any kind of niche opens up in the future, a specific opportunity in a specific industry might be there to automate some task.

Somewhere some entrepreneurial developer will see that and utilize these tools to address it right away, maybe almost instantly at some point.

This isn't general purpose AI that can do anything, but an almost infinite number of specific applications of AI and machine learning used all over the place. When you combine it with things like calc computing, it almost looks like electricity, something that is general purposed and can be leveraged almost anywhere.

DP: The analogy of AI being like electricity or a utility is helpful to visualize the vast changes on the horizon. Another useful model from the book comes from the history of computing. Standard operating systems combined with inexpensive hardware and tools lead to an explosion of software innovation; this was the case with the PC, and later the iPhone, iPad, and Android Apps ecosystems. Can you elaborate on how you see that pattern unfolding with machine learning?

MF: That has been the pattern with PC, Android, iPhone, and the proliferation of apps; I think you will see the same thing with robotics. It is a bit more challenging in robotics because of the hardware component. Hardware components—robotic arms, machine vision systems, and stuff—are getting cheaper and you can already buy kits to build your own robot.

You see companies like Google and Facebook open up their machine learning software. A lot of the components of building machine learning systems are free, or they are getting better and easier to use. Also on the software side, the components or building blocks are getting better. I expect that you are going to see the same kind of phenomenon with infinite numbers of these technologies out there that consume any type of repetitive work.

DP: You mentioned Facebook and Google open sourcing their platforms to developers. I interviewed Chris Anderson for my last book and we talked about his open-source approach to drones with his company 3D Robotics. He makes all patents open-source, and his company benefits from a community of enthusiasts that actively contribute. Linux, Github, and Wikipedia are other examples of open-source projects that drive a tremendous amount of value from a community's contributors.

Do you envision a world where these building blocks are open-source and easily accessible to a community, or will there be a few key players centralizing monopolistic control? How do you see that playing out?

MF: I think to a certain extent, both will exist. What is happening with machine learning, especially deep learning and AI, is pretty focused in companies like Google, Facebook, and Amazon. Those are the big players driving it. But they are making that technology available to utilize in specialized applications. There is a way to tap into and leverage that whether it is open-sourced or proprietary through some kind of revenue share.

I think there will be competing versions of the same things. Some might be open-source and some not. The main point is that there is going to be massive proliferation of this technology and an extraordinary, competitive dynamic going on. Part of what is driving this forward is also that Google is competing with Facebook. We are looking at a huge disruption in terms of where this technology is going.

DP: In my last book Disruption Revolution, *the title referred to a pattern after the economy crashed in 2008–2009, when I noticed people rallying around the term "disruption" even when they hadn't released a product or service. There was also this hype around the TechCrunch Disrupt conference and massive investments in new startups at incredibly high valuations across all industries. I viewed this wave of innovation as a disruption revolution.*

There was this idea that Brian Solis talked about in my book about how constraint drives innovation. When companies are forced to make more utilizing fewer resources (smaller budgets, less staff, etc.), they refocus and prioritize differently. You wrote your first book The Lights in the Tunnel *at the same time. My sense is that you recognized a similar pattern. When jobs were eliminated during the recession, leaders across industries actively looked for solutions to automate rather than replace workers.*

I am curious to what extent the economic crash became a catalyst for automation and the trends you describe. I also wonder if many people are not seeing the troubles that lie ahead because they are still viewing disruption through the lens of economic recovery.

MF: Yes, I think that is definitely true. There is no doubt that recessions are when consolidation happens. Companies don't lay off people for no reason. Very often they lay off because of an economic downturn and they just never hire those people back. They realize that the technology now enables them to avoid rehiring.

Economic evidence is that happens regularly in recessions and jobs that disappear are good, solid, middle-ranged jobs. That doesn't happen gradually. It is focused in recessions. Good jobs disappear, and when we recover, the jobs that come back are these low-wage, burger flipping, Walmart-type of jobs. There is consolidation and a hollow-out effect.

Actually, one of the most innovative periods in U.S. history was the Great Depression in the 1930s. There were lots of technological advances, even though it was an historically apocalyptic time. You can think of that in other ways too. One of the points I make in the book is that we have many people predicting Moore's law will come to at least a temporary plateau. The technology may be reaching its limit. If that's true, I am not sure it's going to be a bad thing.

Moore's law has always been kind of a crutch

Moore's law has always been kind of a crutch, something that everyone expects. Computers are going to get twice as fast, so maybe we don't work as hard on other things as we should. For example, find new ways to hook computers together in parallel architectures, or entirely new architectures and so forth. It wouldn't surprise me if there is a slow-down in Moore's law. That constraint could force people to think in new ways and actually open up new avenues to progress. So yes, I think constraint driving innovation is an important idea.

DP: One of the basic principles of economics going back to Adam Smith is the division of labor—the idea that more specialized skill sets create a more diverse range of products and services, leading to a more diversified and flourishing economy. Yet according to your book, one of the paradoxes of the computer age is that as work becomes more specialized, it may become more susceptible to automation. Can you explain that?

MF: Machines can do specific things. The one tenant of artificial intelligence that everyone will tell you is that we don't have AI that can do general things. What we have is machines running algorithms that can do very specific things. As human work becomes more specific, it gets easier to isolate tasks that people do and then automate them.

I think for the foreseeable future, it is not going to be about building a system that can do everything a particular worker can do. It is going to be about automating a large percentage of the tasks a person does. The things that machines are not able to do will probably get consolidated into fewer jobs. I see that process unfolding for the foreseeable future until we have science-fiction level artificial intelligence that can think at the level of a human being.

DP: I would like to dig deeper into some industry specific examples. When I read stories about robots making burgers, vending machines competing with retail stores, in-store robots or smart assistants delivering added value to my shopping experience, I get excited because this is the kind of future that I dreamed about where the quality of daily life seems to get better and better with these improved efficiencies. Silicon Valley is built on this type of techno optimism.

The flip side is that efficiencies come at the expense of services jobs being cut. Every initiative to raise the minimum wage creates greater incentives for employers to outsource workers and cut costs. Can you tell us about some of the disruptions that you see coming with the retail and hospitality sectors?

MF: I expect a big disruption there. Certainly fast food is going to be ripe with disruption. If you could build a robot that can build an iPhone in China, then you are going to be able to build a robot that can flip hamburgers, make coffee drinks, and all of that stuff. In retail, I think we are going to see robots doing things like stocking retail shelves; we already have robots that can go around and take inventory. There are service robots that will answer questions and take their place in the store to show you where something is.

Probably the most important thing in retail is that people are going to use their mobile device to get information. It's going to be instantly available and very reliable. You will be able to engage with it on a natural language basis and it will probably replace having an employee in the store to answer questions and assist you. People are also using smartphones to pay for products at checkout, so I see that evolving. All of this will mean fewer jobs.

DP: Rise of the Robots *was the* Financial Times *and the* McKinsey *book of the year in 2015. Those are two of the most trusted sources of information for white-collar workers, whose jobs are also at risk of automation. What disruptions to you see on the horizon for office jobs?*

MF: In some ways, it is a lot easier to automate an office job than a fast-food job because you have to build an actual robot for fast food. This is expensive and difficult. You have to deal with cleaning it, and that has a lot of challenges. Automating an office job requires only software.

I saw a recent report that says in the biggest U.S. corporations in corporate financing departments—jobs like accounting, accounts payable, accounts receivable, and financial planning—the head count relative to revenue of the corporation has gone down by about 40% in the last decade. We already see those jobs disappearing. Any kind of job where you are sitting in front of a computer manipulating information in some routine, predictable way can be susceptible to automation. For example, if you are cranking out the same basic report or doing some kind of financial analysis again and again.

There was an article in the *New York Times* about how Goldman Sachs bought a startup company that automated a lot of work done by its analysts. These are the people they hire out of Ivy League colleges. They go to Wall Street and sit in front of spreadsheets for 12 or more hours a day cranking out numbers. A lot of that is highly susceptible. I think this is one of the biggest disruptions we are looking at because these are the jobs that people going to college would like to have. This is not low-wage unskilled work; this is skilled work. I think the impact on these kinds of jobs is one of the biggest challenges we are going to face.

> *This is one of the biggest disruptions...these are the jobs that people going to college would like to have*

DP: In the book, you have an interesting take on outsourcing. Instead of viewing it through the lens of free trade, maybe we should treat it like virtual immigration where workers are entering the country to virtually work. Offshore workers are also the first step towards automation. One study estimated the future impact of offshoring might be 30–40 million U.S. jobs. If a job can be outsourced, then automation would be the next logical step. How might the combination of outsourcing and automation impact the economy?

MF: In many ways and in general, offshoring is the leading edge of automation. It's what you do when you have enough technology to offshore the job, but not enough to completely automate it. That persists for a time but usually not forever. Eventually it moves toward full automation and you see that in many cases. A lot of the jobs that are offshore in countries like India and Philippines right now are being automated.

They are being handled by digital voice technology, call center automation where machines are increasingly answering questions and using technologies like Watson with this powerful natural language capability. You can see how that is going to accelerate rapidly. That will potentially have a huge impact on countries like India that invested heavily in offshoring. They are worried about it. The huge numbers of jobs they created are not going to be around forever.

DP: Some people suggest that education might be the solution to potential widespread unemployment. Learning the right skills enables you to get jobs protected from automation. Some economists have this idea of freeing up, that education empowers people to do more meaningful work when less desirable jobs become automated.

However, in the book you debunk the myth that education could solve the problem of potential widespread unemployment. Why can't we educate our way out of the disruptions that you see?

MF: Education has been the solution historically, and it worked well when machines were taking manual labor jobs, digging ditches and things like that. But now, it has become much more broad-based. Machines are taking skilled jobs, knowledge worker jobs, office jobs. They are moving up the skill level very rapidly, perhaps faster than a person is able to do.

Machines are moving up the skill level very rapidly, perhaps faster than a person is able to do

For example, think of extraordinarily skilled jobs like a radiologist, doctors who look at x-rays, mammograms, and all of that. They went through years of medical school and residency to do that job. Yet machines are getting better and better at it, and I think those jobs will be gone entirely. I suspect that for two reasons: (1) More skilled jobs are being threatened as machines take on cognitive challenges, and (2) There is a limit to how much we can educate people.

If you take that idea to its extreme, then you get to a point where in order to have a job, you have to go to MIT and get a PhD in Artificial Intelligence so that you can be the one developing all of this. You are not going to take the guy flipping hamburgers or stocking shelves and send him to MIT; there are basic limits to what we can train people to do. We may be beginning to encroach on those limits already. I don't think education is ultimately a sustainable solution.

DP: On the subject of sustainability, economic growth is not sustainable if there is widespread unemployment. Market-based economies need large numbers of consumers buying things to keep them running. Wealth consolidation and drastic unemployment could lead the entire economy to contract and create a kind of deflationary spiral. This is what you refer to as the potential for techno-feudalism. Can you elaborate on the scenarios that may arise?

MF: There is a certain level of unpredictability. If you don't have consumers, then you can't have economic growth because you don't have enough demand out there. The market economy or capitalism depends on consumers. Ultimately, consumers drive demand. Without them, you get into that downward spiral or maybe a real economic crisis, not just from people not consuming but also because they can't pay debt. We have been through that before.

There is a range of possibilities. You can get a stagnation situation where the economy doesn't grow or it could be much more severe depending on how widespread unemployment is. In that case, you get into a deflationary spiral, maybe another depression. One of the ideas suggested in the book is if that happens, the economy could somehow adapt to basically siphon off the rich people into their own society where they have their own economy and everyone is kind of left out like in the movie Elysium. That is one scenario, though like I said, it's very unpredictable.

The general thoughts are you need broad-based consumption if you want a successful economy. That is very important. We want to adapt capitalism to ensure that happens.

DP: I think it's important to understand the potentially drastic, worst-case scenarios like you describe in order to emphasize the necessity for radically innovative and new solutions. The consensus among many economists and technologists, including yourself, is that basic income will become necessary to keep broad based consumption going and drive a successful economy.

As we think about widespread disruption to employment and the ubiquitous rise of AI and automation, how would a solution like basic income be implemented?

MF: I think basic income is the best way to solve this problem. People need an income for two reasons. First, so they can survive and are not living on street. Second, they have to be consumers capable of buying the stuff that is produced by the economy.

My vision is that we would introduce a guaranteed basic income at a relatively low level so that it doesn't right away become a disincentive to work. People can still want to work or start a business on top of that. Between this income floor, and whatever else they are doing, they have a sufficient income to survive economically and be active consumers. That helps things thrive.

As the economy continues to grow, technology advances, and automation becomes more prevalent, you would expect to increase that guaranteed income over time. Then maybe people rely less on work and more on this guaranteed income. That seems to be the best way to scale to prosperity across society and continue to have economic growth rather than this stagnation or downward spiral scenario.

DP: We covered a lot of broad ranging topics on the future of employment and the economy. The question probably coming to the minds of many readers is "Ok, now what do I do? And what do we do" What advice would you give people regarding employment and how might we all work together to address the challenges that lie ahead?

MF: There are two things. First, is what you should do as an individual. If I am talking to a college student about careers, some areas are harder to automate such as healthcare, nursing, and so forth. Don't make a big investment in acquiring skills that allow you to do some routine job. What you ought to do instead is emphasize creativity. Do something that involves building something new rather than cranking out the same report. That is the advice to people.

More importantly is the question to us as a society. It is crazy to give advice on how can you as a person succeed in the future if we know that it will not work out for everyone. A lot of people are going to fail and run into real problems. We need to figure out a solution for everyone such as a guaranteed basic income. A big part of what I do in writing and speaking is build awareness because ultimately this is going to require a political response.

It is crazy to give advice on how can you as a person succeed in the future if we know that it will not work out for everyone

We need to get people thinking and talking so this will get on the political radar and we can begin to develop real policies to help us address the problems. But before that can happen, we have to build awareness; we need people thinking about this. That is a big part of what I am trying to do and what I would advise other people to start doing as well.

DP: I love the sense of urgency that you bring to building awareness around the challenges and disruption that lies ahead so that we can co-create the future society that we want to live in.

All of this talk about the challenges and disruption can distract the readers from the simple fact that you write about this stuff because you are incredibly passionate about technology. There are so many exciting things on the horizon like 3D printing, nano technology, autonomous vehicles, and the rise of big data that can create a tremendous amount of value. As a final question: What excites you about the future that lies ahead?

MF: It all excites me! All of these technologies and dimensions have terrific potential. Self-driving cars will save globally hundreds of thousands of lives because they will be safer and we won't have as many accidents. They will be more efficient, greener, better for climate change and real estate utilization in cities. There are tremendous benefits there. But then on the other side, you have the fact that millions of jobs and livelihoods are going to be at risk.

What we want is a scenario where we can leverage the benefits of these technologies, and then also recognize the downside and figure out what to do about it. Otherwise we are headed for big trouble. The biggest impact of 3D printing may be when that scales up to construction and we can 3D print houses and buildings. That has lots of benefits, but globally it also eliminates something on the order of hundreds of millions of jobs. There are huge issues there. We need to leverage the positive benefits of these technologies while addressing the downsides. That is the challenge for us all.

....

MARTIN FORD is a futurist and the author of two books: *The New York Times* best-selling *Rise of the Robots: Technology and the Threat of a Jobless Future* (winner of the 2015 Financial Times/McKinsey Business Book of the Year Award and translated into 19 languages) and *The Lights in the Tunnel: Automation, Accelerating Technology and the Economy of the Future,* as well as the founder of a Silicon Valley-based software development firm. He has over 25 years experience in the fields of computer design and software development. He holds a computer engineering degree from the University of Michigan, Ann Arbor, and a graduate business degree from the University of California, Los Angeles.

He has written about future technology and its implications for publications including *The New York Times, Fortune, Forbes, The Atlantic, The Washington Post, Harvard Business Review,* and *The Financial Times.* He has also appeared on numerous radio and television shows, including NPR and CNBC. Martin is a frequent keynote speaker on the subject of accelerating progress in robotics and artificial intelligence—and what these advances mean for the economy, job market, and society of the future.

....

CONNECTOGRAPHY

Parag Khanna

Parag Khanna's latest book *Connectography* presents a bold vision of an emerging global network civilization connected by clusters of megacities and supply chains. He calls upon us to reject outdated political geographies (nation-states) in favor of a new worldview where digital devices connect us all and urban areas collaborate and compete for resources. Cities are the main drivers of the future.

- We only care about things we are connected to
- Winners in the future will be the ones with new tools
- Connectivity and interdependence shape global identity
- Experiencing the greatest market failure in history

Parag helps connect the dots between macro-level trends like the collaborative economy and autonomous world, and the shift of power that may occur when billions of people in developing nations and emerging markets access the Internet for the first time. This big-picture, overarching global perspective makes Parag an awesome finale to the book. As he says, connectivity is destiny!

DP: Your latest book outlines this idea of an emerging global network civilization and a worldview that looks at geography through the lens of connectivity, which you refer to as "connectography." What do you mean by that?

PK: Connectography is the fusion of connectivity and geography, or what you might call connective cartography. Our understanding of the world privileges political geography (nation-states) whereas we are building an enormous volume of what I would call "functional geography," which is infrastructure that connects societies and cities across borders. Functional geography gives a much better understanding of how human society functions on a day-to-day level.

My motivation is to map connectivity and trace its consequences in a very open-ended way, in terms of who has power in this system, where or what are the drivers of economic growth, what are the consequences in terms of how we relate to the environment, and what are the demographic and migratory ramifications of total connectivity and its moral ramifications.

DP: This book explores themes of empowerment with companies, startups, and global citizens working together to co-create a better future. A key component of that is how the world is being reorganized around cities and the themes of connectivity that you address in Connectography.

Some people refer to this idea of a triple bottom line: people, planet, and profits. What types of values and best practices do you see in this emerging global network civilization? What defines success in this new world order?

PK: We need a sense of collective morality in building a global society. Think of connectivity in terms of things like supply chains and labor rights. For example, a Bangladesh factory collapsed a few years ago and we realized: Hey, wait a minute! All of this clothing is being made for Zara and H&M in this factory in Bangladesh, and it collapsed and killed hundreds of people. Why aren't our major retailers and brands focusing more on their own supply chain?

We only care about things that we are connected to. Part of the problem is that we think governance of relations between countries is structured according to only international treaties. We are not creating the human connections, which happens through the supply chain. For example, where your mobile phone and clothing came from. These connections allow us to feel each other indirectly. The more connections that we have, the greater sense that we are connected. This builds a type of moral or ethical bond between people. Supply chains are crucial vehicles in which we get from us vs. them, from me to we, if you will.

We only care about things that we are connected to

DP: *When I think about connectivity, what comes to mind is an interconnected digital world. For example, I basically live off of Airbnb. I travel almost continuously working as a digital nomad. In the past year, I lived in Thailand, Dubai, Berlin, Macedonia, Belgrade, Budapest, and Boulder. I manage clients via Skype, e-mail, and Voiceover IP. Almost anywhere I can use Airbnb to find an apartment; Uber, Lyft or some on-demand app to get rides; Google Maps, Trip Advisor, and other apps to navigate new cities; and I stay connected to people via social media.*

I feel like a global citizen. Categories like nationality, ethnicity, gender, and religion don't matter as much as shared interests and values. I crisscross the globe from city to city along the lines that you describe, but that is made possible by the digital world. You spoke mostly about connectivity in terms of supply chains and infrastructure. How does the Internet factor into this new world that you describe?

PK: I would reinforce a lot of what you said, not only through my own example but in billions of people interconnected on social media, or using Skype and other platforms to learn languages, forming bonds of identity across geography. Survey data on millennials shows how young people intuitively think pro-connectivity—pro-positive relations with their neighbors, against walls, for environmental sustainability and intergenerational equity, all of these things are the new norms. There is a lot of promise in that. I wholly subscribe to that.

I think the combination of connectivity, generational change, and economic opportunity comes together to reinforce this global digital society. I call it "global social capital" in the book, and I use it as a term to contradict the notion that the most important kind of capital is the local social capital, as portrayed by Robert Putnam or Michael Sandel, when they talk about how a market society is eroding the family bonds and the social foundations of our society. I say, "Wait a minute, that doesn't mean that the alternative is just markets."

Young people intuitively think pro-connectivity—pro-positive relations with their neighbors, against walls, for environmental sustainability

The fact is that technology enabled global social capital. I may be "bowling alone" to use Robert Putnam's phrase and not live in a kind of 19th- or mid-20th-century archetypical social order, but that doesn't mean there is no social capital. Social capital is now global and mediated through technology. To view everything as worse than the "good-old days" and miss the idea of how global social capital forms through connectivity is a gigantic blind spot.

DP: Companies like Google, Facebook, Twitter, Uber, Airbnb, Microsoft, and Apple are the fastest-growing companies in human history. Apple is one example you cite in the book. It has $200 billion in assets, in many ways making it more powerful than many mid-size countries. These companies connect the world. Facebook—when including its products Facebook, Instagram, and WhatsApp—has a population akin to the world's biggest country.

How does the view of the world change when the companies controlling the platforms that connect us have more economic power and reach than countries? What roles and responsibilities should tech companies play in improving the world?

PK: First, it is important to understand that a world of sovereign equality of nations never existed, so it is a fantasy to presume that any shift away from that model represents a threat.

We lived up until very recently in a world of vertically integrated global empires run by Europe. For example, the British East India Company was vastly more powerful than most territories, possessions, and protectorates of any empire, certainly of the British.

Let's be clear about our history first of all so we don't fall into a trap of fantasies about the world that simply isn't true.

The purpose of the state or government is to maximize welfare for its citizens. As far as I am concerned, connectivity is a human right, so whichever outside agent comes in to provide those human rights should be empowered to do so. For example, think of the controversy surrounding how India rejected Facebook's provision of free basic Internet. I am totally on the side of Mark Andreesen, one of Facebook's board members, who is critical of the decision because India has a long history of shooting itself in the foot as an independent country.

As far as I am concerned, connectivity is a human right

The fact is you have hundreds of millions of citizens who have neither physical connectivity, nor additional connectivity, sanitation, nor food for that matter. Yes, there are fears of monopolism, data privacy, and so forth that should be monitored carefully, but governments should allow a company that is very cash-rich on the back of its hardware sales, software, advertising, or whatever the case to be a net provider of connectivity. I worry less about who provides the service than that the service is provided.

There are always power dynamics and tensions over who gets to benefit, but you wouldn't have that problem at all if people weren't connected. We need to be holistic about these kinds of things. People talk about unfair trade, but the problem is too little trade. People worry about the digital divide; the problem is too little Internet access, not an equality of access. There is often this rich world approach to problems that from a utilitarian standpoint is more about too little than too unequal.

DP: We just contrasted countries and companies, and in the book you also contrast countries with megacities. Cities often have more in common with other cities than they do with their surrounding regions. For example, I do a lot of work in Dubai. Many companies succeed building businesses that service the local market, but they struggle to expand into Saudi Arabia or Egypt instead of entering a city like Singapore. How does connectography impact how companies view the world when they think about developing a global strategy?

PK: Megacities relate to the idea that cities not only have more in common with each other than they do with the hinterlands of the countries to which they belong, but cities also think of each other as their biggest markets. Partnerships among cities are the drivers of strategy and in the book I map the future of markets in 50 main megacities. Companies should focus on how human societies organize rather than think of political maps of 200 discrete nations.

DP: We have gone from approximately 1 billion people online in 2005 to around 5 billion today, and that number will continue to rise exponentially until about 2030 when almost everyone on the planet will be connected to the Internet. There are a lot of these green field opportunities for entrepreneurs because there is no legacy infrastructure or incumbents to get in their way.

For example, some of the greatest innovation in mobile payments is happening in Africa, where people basically skipped having a bank account and they now do all of their transactions on their phones. They don't even need physical money to buy stuff locally. How you see the world evolving as billions of people come online for the first time in the upcoming decades?

PK: We can think of this not only in terms of digital infrastructure and leap-frogging, but even in physical infrastructure. For example, in the Northeastern U.S. you could have high-speed rail that people want, but because you have the current Amtrak system there isn't space to build a parallel track and no one wants it in their backyard. Whereas if there were no legacy or incumbent system, you could just build a new one.

The difference between winners and losers in the 21st century is not rich vs. poor or democracy vs. authoritarian but rather old vs. new. Those who have the newer stuff are going to get ahead. Billions of people are coming online and they have newer mobile phones, faster bandwidth speeds, and cheaper access to data. It is remarkable how this is transforming developing countries. They are able to do mobile payments; some are embracing Bitcoin. You can appreciate how competitive they could be from this leap-frogging you describe.

The difference between winners and losers in the 21st century is not rich vs. poor or democracy vs. authoritarian but rather old vs. new

DP: One of the things that I deeply admire about your work is that you are putting forth an ambitious vision for the future and are asking tough questions about how the world should or could be. That is the same spirit of this book. When I look at the model you present, there are so many forces pushing and pulling at such massive scale that it can feel hard to locate agency. How do you influence systems that are so widespread?

PK: There isn't one organized global governance structure to which we all respond and direct our efforts and influence. Definitely life would be a lot easier if that was the case, but that world might not be desirable to live in. In my last book about global governance, I advocated a form of decentralized corporate anarchy, so in many ways I am the last person to go to when it comes to recommendations for how to coordinate efforts that influence global institutions.

I advocate this principle of subsidiaries. Subsidiary is the word for pushing resources and responsibilities to the most local level possible. Ultimately, this means that you have to get your hands dirty if you want to combat exploitation of labor, alleviate poverty, or reduce gas emissions—to wrap your mind and your hands around the supply chain at every step of the way in order to regulate practices. Writing treaties and declarations won't get you anywhere. Climate change is a great example of this.

Personally, I spend a lot of time with mayors at the city level because that is where humanity is. If you want to touch human lives, you need to operate in cities and a lot times with companies. The reality is they are providers of welfare and need to have a certain responsibility in the world, whether it is asset management companies or with governance.

Talk to anyone who will listen—that is the right way to think about how to have influence or impact today. In the media world things are a little different. There are a few key amplifying mouthpieces such as the *New York Times*, *The Daily Mail,* or *China People Daily*. But in the world of actually changing outcomes, there is a lot more conversation to have and a lot more networks to build. Of course, that may be some rationalization of the fact that I am a bit all over the place (laughs).

DP: I spent most of the last 5 years outside the U.S. working with entrepreneurs in various startup ecosystems. One of the things I see is a standardization of infrastructure to support startups in terms of co-working spaces, accelerator programs and incubators, meetups and pitch events, trainings and workshops, etc. At the same time, almost all jobs are becoming on-demand with the rise of the gig economy. How does the rise of the startup ecosystem fit within the context of connectivity? Are we becoming a planet of entrepreneurs?

PK: That would be nice. Yes, there are a lot more entrepreneurs, and I consider anyone making ends meet in a gig economy to be an entrepreneur. We tell our 6-year-old daughter she is an entrepreneur when she is doing her chores. There are obligations or responsibilities to take part in helping out. We want her to think that life is about doing good for the world.

People see gaps all around them because we are presently living through the greatest market failure in human history

People see gaps all around them because we are presently living through the greatest market failure in human history, which is to say that governments are not collectively capable of meeting the needs of their population. United Nations estimates that governments underserve 4 billion people in terms of basic needs and welfare. That is the textbook definition of a market failure. We need to empower anyone willing to fill those gaps and that is what social entrepreneurs, impact investors, and crowd funding does. My previous book *How to Run the World* is all about that.

DP: We have talked a lot about megacities. I am curious what happens to all of the smaller cities and towns along the supply chains and trade routes when we see the rise of the autonomous world, things like drones or self-driving cars, robots, AI, etc.

On the one hand, there is this risk for massive disruption to unemployment. For example, what happens when self-driving trucks become mainstream. In 30 out of 50 U.S. states, the truck driver is the number one job. There are 10 million truck drivers in the U.S. Then there are 10–15 million more people working in hotels, restaurants, gas stations, etc., that service truck drivers. Automation from self-driving trucks could impact 20–25 million jobs in the U.S. alone.

On the other hand, the biggest cost of trucking is the driver. Eliminating labor would cut shipping costs by as much as 70% and since self-driving trucks don't need to sleep, they could be 50% faster. Self-driving trucks could enable people to work from anywhere, receiving supplies and allowing them to pursue utopian visions of community with solar power and energy efficient batteries, vertical farming, and living off the grid.

How do you see things changing outside of cities?

PK: There are a lot of countervailing trends. Many people say things like, "Hey, why do we need all of this urbanization? You can just plant a fiber cable out to a farm and let people work remotely in their blue collar country-side home, tend to their livestock and cattle, and have their cappuccino while online trading Bitcoins and checking marketplaces for commodities."

That is not the way world really works. We are not going to plant fiber cables out to the most distant farms and people don't move to cities just so they can have fiber cables. People are in search of digitization of things like healthcare and jobs. We will never reverse urbanization. People will always move to cities.

We should connect remote geographic areas through better roads, highways, and so forth. For example, in the *New York Times* I talked about the Appalachia region. We declared war on poverty 50 years ago in Appalachia and didn't make a real dent in it because we didn't build good roads, highways, and railways. It would have been smarter to do that over the last 50 years than to wage a so-called war on poverty.

When it comes to the technology, of course there is going to be labor displacement as a result of things like driverless trucks. The question becomes: Do we want to be a nation of truck drivers? If you don't want to be a nation of truck drivers, then let the driverless trucks come along and train truck drivers to be mechanics. Give them jobs in an advanced automobile manufacturing hub, which is what Tennessee and Kentucky should become rather than fighting for low-wage jobs in Korean automobile plants.

We can take advantage of technological disruptions and transplant people into new and higher wage industries. I point out endlessly in the book that countries like Germany and South Korea have vocational education systems that focus on apprenticeship. That is how they maintain low unemployment and keep wages growing in higher end sectors of the economy. The U.S. isn't doing that, and it is our own fault if we don't do so.

We can take advantage of technological disruptions and transplant people into new and higher wage industries

Health care and education are non-trainable sectors that technology doesn't really displace. Robots don't teach our kids; robots aren't doctors. This is the fastest growing area in the U.S. economy, but it could grow a lot more if we fixed the mismatch between what we train people to do and what the needs are. People want to do those jobs, but they are not educated in the right way. We can fix these problems and have been ignoring them for more than a decade

DP: I did Ph.D. studies in Religions of the Americas at Princeton. One of the things that I studied was the history of colonialism and the rise of fundamentalism in the modern world.

There is this way in which fundamentalist or extreme views tend to pop up at the boundaries between rural and urban areas. Often times people on the fringes are marginalized and they struggle economically due to disruption in traditional industries like agriculture and manufacturing. Groups that are "different" such as immigrants or refugees are often blamed for problems like loss of jobs or rises in crime that result from unemployment. All of the disruption can lead to an emphasis of identities anchored in religion, nationalism, or tradition.

I think of that in contrast to people in major megacities. They view the world through economics, cooperation, and shared value created by living together. They embrace collaboration and sharing, invest in education and the commons, and the types of things that we talked about. It seems like the polarizing extremism of contemporary identity politics could be viewed through the lens of connectivity. How does your model relate to identity in the modern world?

PK: Global identity cannot just be this wishful aspirational thing. It emerges from the process of building more connectivity and interdependencies between people economically, digitally, and socially. That sense of global identity amplifies by generational change, in which a cohort of people views this as natural to them because it's the way things have always been. It's a process that happens not simply through infrastructural connectivity. I hope that we will not unwind or reverse it because in some ways it's just getting started now.

DP: I love how you keep bringing conversations about ideas and big picture vision back to pragmatic things like infrastructure and connection that are very tangible. As a final question: What advice would you give in terms of practical steps on how to succeed in this new emerging world?

PK: Be connected. Be mobile. For example, be willing to transplant geography from failing cities to successful cities, or from slower domestic markets to more thriving international markets. Be resilient and capture these trends. That is how you prepare to stay afloat and succeed.

In terms of the betterment of humankind, there are all types of things we could do right if we built more connectivity, but we are not there yet. We talked earlier about technology companies being important agents of connectivity. I want to see a lot more of that.

We want more peer-to-peer capitalism, which I call the global digital workforce. There is a strong case for Internet penetration as an agent of economic growth. This can also boost the services share of economy. Societies have to transition over to services more rapidly because of what Martin Ford writes about, who you also interviewed for this book.

Be connected. Be mobile...Be resilient and capture these trends. That is how you prepare to stay afloat and succeed

If we want a world built more around the sharing economy and peer-to-peer capitalism, it is going to rest on the backbone of digital infrastructure, supply chains, and all of the things we spoke about earlier. There are reinforcing dynamics between what we are doing in physical infrastructure, which we take entirely for granted, and what our goals are for human society and for our economy. It's all interconnected.

PARAG KHANNA is a leading global strategist, world traveler, and best-selling author. He is a Senior Research Fellow in the Centre on Asia and Globalisation at the Lee Kuan Yew School of Public Policy at the National University of Singapore. He is also the Managing Partner of Hybrid Reality, a boutique geostrategic advisory firm, and Co-Founder & CEO of Factotum, a leading content branding agency.

Parag's latest book is *Connectography: Mapping the Future of Global Civilization* (2016). He is also co-author of *Hybrid Reality: Thriving in the Emerging Human-Technology Civilization* (2012) and author of *How to Run the World: Charting a Course to the Next Renaissance* (2011) and *The Second World: Empires and Influence in the New Global Order* (2008). In 2008, Parag was named one of Esquire's "75 Most Influential People of the 21st Century," and featured in WIRED magazine's "Smart List." He holds a PhD from the London School of Economics, and Bachelors and Masters degrees from the School of Foreign Service at Georgetown University. He has traveled to more than 100 countries and is a Young Global Leader of the World Economic Forum.